T0291145

Mobile Robots for Digital Farming

This book provides a complete and comprehensive reference for agricultural mobile robots, covering all aspects of the design process, from sensing and perceiving to planning and acting for practical farming applications.

Mobile Robots for Digital Farming explores topics such as Robot Operating Systems (ROS), dynamic simulation, artificial intelligence, image processing, and machine learning. Additionally, it features multiple case studies from funded projects and real-field trials.

This book will be useful for professors and academics in various engineering disciplines (mechanical, robotics, control, electrical, computer, and agricultural), graduate and undergraduate students, farmers, commercial growers, startups, private companies, consultancy agencies, equipment suppliers, and agricultural policymakers.

Redmond R. Shamshiri is a scientist at the Leibniz Institute for Agricultural Engineering and Bioeconomy (ATB) working toward digitization of agriculture for food security. He holds a Ph.D. in agricultural automation with a focus on control systems and dynamics. His research fields include autonomous navigation of agricultural mobile robots, sensor fusion and artificial intelligence for collision avoidance, Teleoperation, wireless systems, and optimization of controlled environment agriculture.

Ibrahim A. Hameed is currently Professor and Deputy Head of research and innovation, and Head of the international master program in simulation and visualization with the Faculty of Information Technology and Electrical Engineering, Norwegian University of Science and Technology, Trondheim, Norway, where he has been Associate Professor with the Department of ICT and Natural Sciences, since 2015.

Mobile Robots for Digital Farming

Edited by Redmond R. Shamshiri and
Ibrahim A. Hameed

CRC Press

Taylor & Francis Group
Boca Raton London New York

CRC Press is an imprint of the
Taylor & Francis Group, an **informa** business

First edition published 2025
by CRC Press
2385 NW Executive Center Drive, Suite 320, Boca Raton FL 33431

and by CRC Press
4 Park Square, Milton Park, Abingdon, Oxon, OX14 4RN

CRC Press is an imprint of Taylor & Francis Group, LLC

ISBN: 978-1-032-30466-3 (hbk)
ISBN: 978-1-032-30692-6 (pbk)
ISBN: 978-1-003-30628-3 (ebk)

DOI: 10.1201/9781003306283

Typeset in Times
by Apex CoVantage, LLC

To Professor Warren Dixon of the University of Florida, with heartfelt appreciation for his enlightening lectures, and special thanks to Maryam Behjati for her invaluable support and encouragement.

Contents

About the Editors xi
List of Contributors xiii

Chapter 1   Sensors, Algorithms, and Software for Autonomous
                  Navigation of Agricultural Mobile Robots 1

Redmond R. Shamshiri

1.1   An Overview of Agricultural Mobile Robots 1
1.2   The Sense, Think, Act Paradigm 3
1.3   Sensors for Autonomous Navigation and
        Obstacle Detection 4
        1.3.1   Sensors for Localization and Mapping..................... 6
        1.3.2   Sensors for Simultaneous Localization and
                  Mapping (SLAM).. 10
        1.3.3   Sensors for Obstacle Detection............................... 12
1.4   Algorithms for Autonomous Navigation and Collision
        Avoidance 19
        1.4.1   Localizations and Mapping 20
        1.4.2   Path Planning... 24
        1.4.3   Path Tracking... 24
        1.4.4   Trajectory Planning 25
        1.4.5   Sensor Fusion Algorithms 25
        1.4.6   Collision Avoidance Algorithms 26
        1.4.7   Machine Learning 27
1.5   The Robot Operating System (ROS) Packages 28
1.6   Simulation Software 32
1.7   Technological Challenges Toward Commercialization........ 36
1.8   Conclusion 36
References 37

Chapter 2   Robot-Assisted Soil Apparent Electrical Conductivity
                  Measurements in Orchards 55

*Dimitrios Chatziparaschis, Elia Scudiero, and
Konstantinos Karydis*

2.1   Introduction 55
2.2   System Design and Integration of Key Components 58
        2.2.1   Soil Conductivity and Employed Sensor 58
        2.2.2   Mobile Robot Setup... 59
        2.2.3   Positioning System Integration for Field
                  Navigation... 60

          2.2.4   Robot Configuration and the Design of the
                  Sensor Mounting Platform 61
    2.3   Development and Calibration for Optimal
          Sensor Placement 62
          2.3.1   Determination of Robot Interference in Soil
                  Conductivity Measurements 63
          2.3.2   Evaluation of Robot Platform Maneuverability
                  Though Gazebo Simulation 68
          2.3.3   Preliminary Feasibility Experimental Testing
                  of Boundary Configurations 72
    2.4   Field-Scale Experiments 76
    2.5   Conclusion 80
    References 85

Chapter 3   Electrical Tractors for Autonomous Farming 89

          Redmond R. Shamshiri

    3.1   Introduction 89
    3.2   Background and Applications 90
    3.3   Benefits of Using E-Tractors 91
    3.4   Challenges and Limitations with E-Tractors 93
    3.5   Availability of E-Tractors 94
    3.6   E-Tractors for High-Density Orchards 98
    3.7   Accelerating the Adoption of E-Tractors 103
    3.8   Conclusion 104
    References 105

Chapter 4   Agricultural Robotics to Revolutionize Farming:
          Requirements and Challenges 107

          *Redmond R. Shamshiri, Eduardo Navas, Jana Käthner, Nora
          Höfner, Karuna Koch, Volker Dworak, Ibrahim Hameed,
          Dimitrios S. Paraforos, Roemi Fernández, and Cornelia Weltzien*

    4.1   Introduction 107
    4.2   Advances in Robotic Manipulators for Agriculture 109
          4.2.1   Advances in Visual Servoing and
                  Computer Vision 110
          4.2.2   Advances in Robotic Pruning, Thinning, and
                  Trimming 112
          4.2.3   Advances in Robotic Weeding and
                  Target Spraying 114
          4.2.4   Advances in Robotic Harvesting 115
    4.3   Human-Robot Collaboration 118
    4.4   Advances in Soft Robotics and Soft Grippers 119
          4.4.1   Soft Robotic Manipulators 119

4.4.2   Innovations in Soft Grippers 121
4.4.3   Finger-Tracking Gloves 122
4.5   Advances in Field Robots and Their Availability
       in Europe 124
4.6   A Case Study for Potato Fields 125
4.7   Current Challenges and Limitations of
       Agricultural Robotics 139
4.8   Future Scenarios 142
4.9   Conclusion 144
4.10   Acknowledgment 145
4.11   Disclaimer 145
References 145

Chapter 5   Toward Optimizing Path Tracking of Agricultural Mobile
Robots with Different Steering Mechanisms: A Simulation
Framework 156

Redmond R. Shamshiri

5.1   Introduction 156
5.2   Effect of Steering Mechanism on Path Tracking 157
5.3   Finding the Shortest Path 160
5.4   Path Tracking Controllers 164
       5.4.1 PID Controller 165
       5.4.2 Model Predictive Controller 168
5.5   Conclusion 169
References 171

Index 185

About the Editors

Dr. Redmond R. Shamshiri is a scientist at the Leibniz Institute for Agricultural Engineering and Bioeconomy (ATB) working toward digitization of agriculture for food security. He holds a Ph.D. in agricultural automation with a focus on control systems and dynamics. His research fields include autonomous navigation of agricultural mobile robots, sensor fusion and artificial intelligence for collision avoidance, Teleoperation, wireless systems, and optimization of controlled environment agriculture.

Dr. Ibrahim A. Hameed is currently Professor and Deputy Head of research and innovation, and Head of the international master program in simulation and visualization with the Faculty of Information Technology and Electrical Engineering, Norwegian University of Science and Technology, Trondheim, Norway, where he has been Associate Professor with the Department of ICT and Natural Sciences, since 2015.

Contributors

Dimitrios Chatziparaschis
Dept. of Electrical and Computer
Engineering
University of California
Riverside, CA

Volker Dworak
Leibniz Institute for Agricultural
Engineering and Bioeconomy (ATB)
Potsdam, Germany

Roemi Fernández
Centre for Automation and Robotics,
UPM-CSIC, Carretera CAMPO-
REAL, Arganda del Rey
Madrid, Spain

Ibrahim Hameed
Department of ICT and Natural
Sciences, Faculty of Information
Technology and Electrical
Engineering
Norwegian University of Science and
Technology (NTNU)
Ålesund, Norway

Nora Höfner
Leibniz Institute for Agricultural
Engineering and Bioeconomy (ATB)
Potsdam, Germany

Konstantinos Karydis
Dept. of Electrical and Computer
Engineering, University of California
Riverside, CA

Jana Käthner
Leibniz Institute for Agricultural
Engineering and Bioeconomy (ATB)
Potsdam, Germany

Karuna Koch
Leibniz Institute for Agricultural
Engineering and Bioeconomy (ATB)
Potsdam, Germany

Eduardo Navas
Leibniz Institute for Agricultural
Engineering and Bioeconomy (ATB)
Potsdam, Germany

Dimitrios S. Paraforos
Institute of Agricultural Engineering,
University of Hohenheim
Germany

Elia Scudiero
Dept. of Environmental Sciences,
University of California
Riverside, CA

Redmond R. Shamshiri
Leibniz Institute for Agricultural
Engineering and Bioeconomy
(ATB)
Potsdam, Germany

Cornelia Weltzien
Leibniz Institute for Agricultural
Engineering and Bioeconomy (ATB)
Potsdam, Germany

1 Sensors, Algorithms, and Software for Autonomous Navigation of Agricultural Mobile Robots

Redmond R. Shamshiri

1.1 AN OVERVIEW OF AGRICULTURAL MOBILE ROBOTS

Over the last two decades, and with the growing demand for efficient and cost-effective farming practices [1, 2], agricultural mobile robots with advanced sensing and navigation systems have emerged to reduce the human workforce for repetitive tasks in harsh field conditions. These robots are facing a breakthrough evolution led by research and development in autonomous navigation and collision avoidance technologies. Robotization is part of the larger trend of digitalization in agriculture [3], which involves the use of technology to improve the efficiency, productivity, and sustainability of farming practices such as planting [4], spraying [5], harvesting [6], and crop monitoring [7]. From the earliest prototypes of automated tractors that dates back to the late 1960s [8] to the advanced robotic platforms of today, a wide variety of components [9], including sensors, algorithms, and controllers, have been evaluated for navigation [10], obstacle detection [11], and collision avoidance [12] to complete farming operation in a safe manner and with minimal human intervention.

Autonomous navigation refers to a robot's ability to determine its location and plan a path to a target. The early robots developed for farming were limited in their technology and were mainly used for plowing and planting [13–15]. Today, a wide range of mobile robots is available from different companies, each designed to perform specific tasks and meet the needs of different types of agricultural fields. They benefit from a series of sensors, algorithms, data communication, and software integration for positioning and mapping to navigate through complex and unpredictable fields and orchards without causing damage to crops, animals, equipment, or humans. Figure 1.1 shows examples of such robotic platforms, including (a) AGXEED AgBot [16], (b) Thorvald [17], (c) SMASH [18] (produced by a project called Smart Machine for Agricultural Solutions Hightech in cooperation with 10 technology partners), and (d) Swarmbot 5 [19]. The AGXEED AgBot is an autonomous robot equipped with

DOI: 10.1201/9781003306283-1

Real-Time Kinematic (RTK) Global Navigation Satellite Systems (GNSS) and collision avoidance sensors for precise guidance and safe positioning in soil preparation,
seedbed preparation, and seeding. Thorvald is a versatile robot designed to work in
a variety of weather conditions and terrains. SMASH is a modular mobile robot that
can monitor and control crops, collect soil samples for analysis, and accurately target
agricultural chemicals for precision application. Swarmbot 5 from SwarmFarming
is a new paradigm for farm applications where swarms of small, agile, autonomous
platforms can perform mowing on turf farms and between orchard rows. A list of
some of the active companies in producing agricultural mobile robots is provided in
Table 1.1.

FIGURE 1.1   Examples of agriculture mobile robots, showing (a) AGXEED AgBot [16], (b)
Thorvald [17], (c) SMASH [18], and (d) Swarmbot 5 [19].

TABLE 1.1
Some of the Active Companies in Producing Agricultural Mobile Robots

Company Name	Main Field(s) of Activity	Produced Robot(s)
Blue River Technology (John Deere) www.bluerivertechnology.com	Uses computer vision and machine learning to target and spray individual weeds in crop fields	See & Spray
Naïo Technologies www.naio-technologies.com	Specializes in agricultural robots designed to assist with tasks such as weeding, inter-row soil management, and crop monitoring	Oz, Dino, Orio, Jo
Clearpath Robotics www.clearpathrobotics.com	Provides rugged mobile robot platforms like the Husky and Warthog, which can be customized for various agricultural applications	Clearpath Husky, Clearpath Warthog
Agrobot www.agrobot.com	Specializes in robotic solutions for strawberry harvesting, with robots designed to pick ripe strawberries without damaging the plants	Agrobot Strawberry Robot
SeedMaster Manufacturing www.seedmaster.ca	Develops an autonomous robotic system for various agricultural tasks, including seeding and spraying	Dot Power Platform

TABLE 1.1 *(Continued)*
Some of the Active Companies in Producing Agricultural Mobile Robots

Company Name	Main Field(s) of Activity	Produced Robot(s)
FarmWise www.farmwise.io	Develops AI-powered robots for precision weeding and crop care, with a focus on reducing the use of chemicals and increasing farm efficiency	FarmWise Robot
Yanmar www.yanmar.com	A manufacturer of agricultural machinery and autonomous tractors designed to enhance precision agriculture and improve farm productivity	Harvest Automation Robot
SwarmFarm Robotics www.swarmfarm.com	Offers autonomous agricultural robots for tasks such as precision planting, weeding, and crop monitoring, promoting sustainable farming practices	SwarmFarm Robot
CNH Industrial www.cnhindustrial.com	Develops autonomous tractors and agricultural machinery, including Case IH and New Holland, focusing on advancing farm automation	Autonomous Tractor
Octinion www.octinion.com	Is known for Rubion, an autonomous strawberry-picking robot that uses computer vision and soft-touch technology for delicate fruit harvesting	Rubion Strawberry Harvester
ecoRobotix www.agravis-robotik.de	Focuses on sustainable farming by providing weeding robots that use computer vision to detect and eliminate weeds in row crops, reducing herbicide use	ecoRobotix Weeding Robot

1.2 THE SENSE, THINK, ACT PARADIGM

The multidisciplinary field of autonomous mobile robots is usually summarized through the fundamental concept of 'Sense, Think, Act' [20, 21], as shown in Figure 1.2. This paradigm captures the essence of how robots perceive their environment, make decisions, and execute actions to navigate and interact with the world around them. At the heart of autonomous navigation is the ability of a robot to localize itself and sense its surroundings. This involves gathering information about its coordinate location and the environment (i.e., providing the robot with data on its position, orientation, nearby objects, terrain, and other relevant information) using various sensors [22], most commonly GPS, IMUs (Inertial Measurement Units), cameras, light detection and ranging (LIDAR), and distance detection sensors, creating the foundation for decision-making and planning. The cognitive aspect of autonomous navigation involves processing the sensory data, interpreting it to understand the current situation, and then formulating a course of action. Algorithms and AI techniques, such as machine learning, path planning, and control theory, are employed to navigate the robot through complex scenarios. The 'thinking' phase includes tasks such as route planning [23], obstacle avoidance, trajectory generation [24], and high-level decision-making based on goals or mission objectives. This decision-making process is achieved using heuristic or optimal methods [25] and involves assessing risk, predicting future states, and optimizing actions for efficiency and safety. These approaches represent different philosophies and methodologies for guiding robots through their environment. The final stage of autonomous navigation is putting the decisions into action. The robot translates its planned actions into physical movements or control commands, which involves actuating the robot's motors, wheels, or other mechanisms to move through its environment. It encompasses not only motion control [26] but also coordination among different subsystems, such as propulsion, steering, braking, and communication. It should be noted that the robot continuously monitors its own state and adapts its actions in real time to ensure that it remains on course and responds to any unexpected changes in the environment. Additionally, the 'act' phase may involve interaction with objects or humans, such as picking up fruits [27], planting seeds [4], or removing weeds [28].

1.3 SENSORS FOR AUTONOMOUS NAVIGATION AND OBSTACLE DETECTION

The effective operation of mobile robots in complex and dynamic agricultural environments is not possible without a sophisticated and reliable set of sensors to constantly deliver perception data. GNSS, including GPS with enhanced techniques such as RTK and Differential GPS (DGPS), are foundational for global localization [29, 30], providing precise position information, enabling robots to navigate

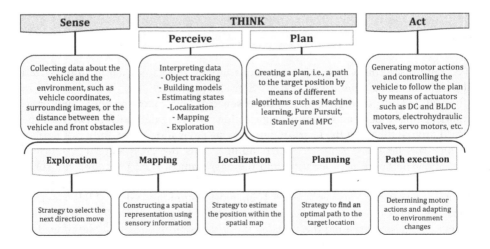

FIGURE 1.2 An overview of the Sense, Think, Act paradigm for autonomous navigation of agricultural mobile robots.

large fields and orchards with accuracy. Inertial Measurement Units (IMUs) are angular sensors that provide data on orientation and angular velocity and are usually integrated with GPS data to enhance the robot's ability to maintain its course and accurately position itself within the field [31]. Dead reckoning techniques [32] are used for estimating a robot's position based on its previous known position, orientation, and motion [33] to help maintain position accuracy when IMU measurements are suffering from low sensitivity and noise, or when GNSS signals are temporarily disrupted, such as when operating in densely vegetated areas [34].

In the last two decades, vision-based sensors [35], such as high-resolution RGB cameras [36] with a wide choice of advanced computer vision algorithms [37], have undergone significant advancements, transforming how robots perceive their environment. Thermal cameras [38] detect variations in temperature, contributing to safety features by identifying the presence of humans, animals, or machinery in the robot's surrounding and preventing collisions with personnel or livestock. Infrared (IR) cameras [39], as well as multi-spectral and hyperspectral sensors [40], capture information beyond the visible spectrum (i.e., night vision) and enable detailed analysis of obstacles in the field. Depending on the interest range of detecting distance, Radio Detection and Ranging (RADAR) sensors employ radio waves that are effective in adverse weather conditions and can penetrate foliage, making them valuable in dense agricultural environments for providing accurate distance and speed measurements, enhancing ambient awareness, detecting obstacles, and avoiding collisions [41]. Laser-based sensors and three-dimensional LiDAR (Light Detection and Ranging) systems, such as Velodyne VLP-16 and Hokuyo URG-04LX, have become staples in agricultural robot design to create detailed 2D and 3D maps of the

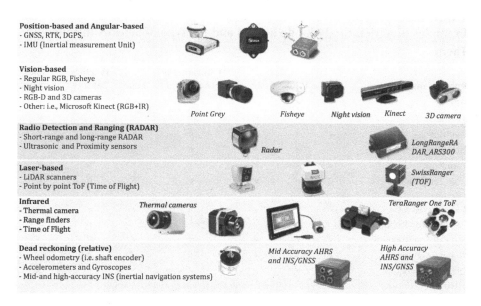

FIGURE 1.3 Different types of sensors used for autonomous navigation of tractors and agricultural mobile robots.

surroundings, offering high-resolution and accuracy that are valuable for obstacle detection and precise mapping [10]. An overview of some of the sensor candidates that can be used in the autonomous navigation of mobile robots is presented in Figure 1.3, followed by a list of some well-known agricultural robotic platforms in Table 1.2 and the types of sensors installed on them.

1.3.1 SENSORS FOR LOCALIZATION AND MAPPING

The core of autonomous navigation of mobile robots relies on local and global localization and mapping [42] using a diverse range of mapping sensors [43] that often include LiDAR systems, which leverage laser beams to capture precise spatial data, along with RGB camera and other vision-based perception systems that provide visual input for environmental modeling. Additionally, multi-channel IR and Time-of-Flight (ToF) laser [44] and ultrasonic sensors contribute to mapping and localization by offering three-dimensional representations of the robot's surroundings. Global localization of mobile robots revolves around the precise determination and maintenance of their position and orientation within their environments, and heavily relies on data from GPS and IMU modules. Table 1.3 presents a list of the most commonly used GPS receivers (along with a summary of their technical specifications, accuracy, communication interface, and price range) that have been reported in the published literature for the autonomous navigation of tractors, mobile robots, and other agricultural machinery.

TABLE 1.2
Examples of Agricultural Mobile Robots and the Type of Sensors Installed on Them

Robot Name	GPS	IMU	Cameras	Ultrasonic or IR Sensors	Lidar	Weather Sensors	Soil Sensors
AgBot	✓	✓	✓	×	✓	×	✓
Thorvald	✓	✓	✓	×	✓	✓	✓
BoniRob	✓	×	✓	✓	✓	×	✓
Blue River See & Spray	✓	✓	✓	✓	×	×	×
FarmWise	✓	✓	✓	×	×	×	✓
Naio Technologies (OZ Robot)	✓	✓	✓	✓	×	×	✓
Yanmar (Harvest Automation Robot)	✓	✓	✓	×	✓	×	×
SwarmFarm Robotics	✓	✓	✓	✓	✓	×	×
Kubota Agrirobo	✓	×	✓	✓	✓	×	×
Clearpath Robotics (Husky)	✓	✓	✓	✓	×	×	×
CNH Industrial Autonomous Tractor	✓	✓	✓	×	×	×	×

TABLE 1.3

List of Most Commonly Used GPS Receivers for Autonomous Navigation of Tractors and Agricultural Mobile Robots

GPS Receiver Model	Summary of Technical Specification	Accuracy	Interface	Price Range
NovAtel OEM7 GNSS [45]	Multi-constellation support (GPS, GLONASS, Galileo, BeiDou). High-precision positioning. Scalable performance options. Robust interference mitigation and GNSS+INS integration.	Cm-level	Ethernet, Serial (RS-232), CANBUS, USB	$5,000–$10,000
u-blox ZED-F9P GNSS [46]	Multi-band and multi-constellation support. Real-time kinematic (RTK) positioning capability. High update rate. Low power consumption and small form factor.	Cm-level	UART (Serial), USB, I2C, SPI, Ethernet	$200–$400
Hemisphere GNSS XF2 [47]	Multi-constellation and multi-frequency GNSS support. Scalable accuracy options. Advanced anti-jamming and interference mitigation. Compact and rugged design.	Sub-meter to Cm-level	Ethernet, Serial (RS-232), CANBUS, USB	$1,000–$3,000
Trimble AgGPS 372 GNSS [48]	Dual-antenna capable for precise heading determination. OmniSTAR correction support. High update rate. Robust GNSS tracking and correction options.	Sub-meter to Cm-level	Serial (RS-232), CANBUS, USB	$2,000–$5,000
Leica Geosystems GS18 T [49]	Self-leveling GNSS receiver with IMU integration. Multi-constellation and multi-frequency support. Fast initialization and high-accuracy positioning. Ideal for mobile mapping and robotics applications.	Cm-level	Serial (RS-232), Ethernet, USB	$15,000–$20,000
Topcon HiPer SR GNSS [50]	Compact and lightweight design. Dual-frequency L1/L2 GNSS tracking. High update rate. Robust anti-theft features and security options.	Sub-meter to Cm-level	Serial (RS-232), USB, Bluetooth, LAN	$2,500–$4,000
Ag Leader GPS 6500 [51]	Dual-frequency GPS and GLONASS support. Real-time kinematic (RTK) correction compatibility. Compact and durable design. Integration with Ag Leader's precision farming solutions.	Sub-meter to Cm-level	Serial (RS-232), CANBUS, USB	$2,500–$4,000

Table 1.4 provides an overview of commercially available GPS navigation toolboxes commonly used in tractors and agricultural mobile robots. It summarizes their communication interfaces (such as CANBUS, ISOBUS, Bluetooth, and Ethernet) and signal correction methods, which are essential for improving the reliability of GPS-based navigation and reducing positional errors, even in areas with obstructed satellite views or atmospheric interference. These signal corrections methods include RTK, D-GPS, Precision Point Positioning (PPP), Satellite-Based Augmentation Systems (SBAS), Wide Area Augmentation System (WAAS), European Geostationary Navigation Overlay Service (EGNOS), and Global Navigation Satellite System (GLONASS). For those robots that have various agricultural implements and equipment, CANBUS and ISOBUS provide standard communication between the navigation toolbox and the control units of the implements, making possible the exchange of critical data, commands, and feedback [52].

One of the most widely used toolboxes for autonomous navigation is the John Deere StarFire GPS Systems [53] which are compatible with John Deere machinery and are well-known for their precision and reliability. The Case IH AFS AccuGuide system [54] also offers high-precision guidance and steering solutions, ensuring accurate paths for tractors and mobile robots. Ag Leader SteerCommand and InCommand systems [55] provide a combination of advanced guidance, steering capabilities, and comprehensive data management, enhancing reliability for the navigation of mobile robots. Raven Industries Viper 4 and Slingshot systems [56, 57] contribute real-time guidance and data communication for remote monitoring within fields. The Topcon X14, X25, and AGI-4 systems [58] offer a suite of precision solutions, encompassing advanced GPS guidance and high-accuracy steering with signal correction [59] for driverless tractors and mobile robots. These commercially available toolboxes represent a diverse array of solutions, each accurately custom-made to address GPS-based autonomous navigation under different field conditions and for various applications, such as precision planting, spraying, weeding, and mowing.

Although D-GPS and RTK-GPS work well in some open-field environments, they still encounter limitations and signal obstructions in other GPS denied fields and orchards, such as the ones shown in Figure 1.4 [44, 60]. The problems associated with poor GPS signal (i.e., loss of connectivity and inconsistent or disrupted signal) can cause inaccurate navigation, loss of task performance, and increased risk of damage to the plants and the robot itself. To assist the navigation system, autonomous mobile robots increasingly adopt local localization techniques [61] by using LiDARs, cameras, wheel encoders, sensor fusion strategies [62], and visual methods that combine data from multiple sensors to enhance accuracy and robustness. These strategies include image-based localization using cameras for visual odometry [63], or capturing visual features such as landmarks or environmental patterns that can be matched to a reference map to estimate the robot's position and orientation [64]. LiDAR-based approaches [10] play a significant role in generating high-definition maps of the environment and comparing real-time LiDAR data with reference maps. Once the robot establishes its location, the subsequent step involves path planning [65] for generating a safe and efficient route from the robot's current position to

TABLE 1.4
Commercially Available GPS Navigation Toolboxes for Autonomous Navigation of Tractors and Agricultural Mobile Robots

GPS Navigation System	Communication/Interface				Signal Correction
	CANBUS	ISOBUS	Bluetooth	Ethernet	
John Deere StarFire GPS Systems	✗	✓	✓	✓	StarFire, SF3, SF1
Case IH AFS AccuGuide	✗	✓	✓	✓	RTK, WAAS, EGNOS, Glonass
Ag Leader SteerCommand, InCommand	✓	✗	✓	✓	RTK, WAAS, EGNOS, Glonass
Raven Industries Viper 4, Slingshot	✓	✗	✓	✓	RTK, SBAS, Glonass
Topcon X14, X25, AGI-4	✓	✗	✓	✓	RTK, SBAS, Glonass
TeeJet Technologies Matrix Pro, FieldPilot	✓	✗	✓	✓	RTK, SBAS, Glonass
NovAtel Various GNSS Receivers	✓	✗	✗	✓	RTK, SBAS, Glonass
Hemisphere GNSS Crescent, Vector, R330	✓	✗	✗	✓	RTK, SBAS, Glonass
AgJunction Outback MAX, eDriveX	✓	✗	✓	✓	RTK, SBAS, Glonass
Leica mojoMINI, mojo3D, GeoSteer	✓	✓	✓	✓	RTK, SBAS, Glonass
Kubota Precision Farming	✓	✗	✓	✓	RTK, SBAS, Glonass
Precision Planting 20/20 SeedSense	✓	✗	✓	✓	RTK, SBAS, Glonass
Farmobile PUC, DataEngine, DataStore	✓	✗	✓	✓	N/A (Data Collection)
Outback Guidance eDriveTC, eDriveXC	✓	✗	✓	✓	RTK, SBAS, Glonass
Case IH AFS AccuGuide	✗	✓	✓	✓	RTK, WAAS, EGNOS, Glonass
John Deere AutoTrac	✗	✓	✓	✓	RTK, SF3, SF1
Trimble CenterPoint RTX	✓	✓	✓	✓	RTK, CenterPoint RTX
Ag Leader SteerCommand	✓	✗	✓	✓	RTK, WAAS, EGNOS, Glonass
Raven Industries SmarTrax MD	✓	✗	✓	✓	RTK, SBAS, Glonass
Topcon Agriculture System 350	✓	✗	✓	✓	RTK, SBAS, Glonass

FIGURE 1.4   View of different orchards with poor GPS signals where mapping, localization, and path planning is crucial for autonomous navigation and collision avoidance [44].

its intended destination while avoiding obstacles. This necessitates consideration of the robot's physical constraints, the layout of the environment, and the presence of dynamic obstacles.

1.3.2   Sensors for Simultaneous Localization and Mapping (SLAM)

Simultaneous Localization and Mapping (SLAM) [42] is an algorithmic approach that enables a robot to build maps of its surroundings in unfamiliar and unstructured environments while concurrently determining its precise position within these maps. This offers a powerful tool for agricultural mobile robots to autonomously navigate and operate within complex and ever-changing agricultural fields, orchards, and vineyards. SLAM relies on a network of sensors, most commonly LiDARs, which provide accurate and high-resolution depth information, facilitating the creation of detailed 2D or 3D maps. By emitting laser beams and precisely measuring their return times, LiDARs generate a dense point cloud [66], representing the spatial layout of objects and obstacles in their vicinity. This point cloud data serves as the foundation upon which SLAM algorithms build detailed maps of the agricultural terrain. The 360-degree scanning capabilities of LiDARs and the panoramic perspective allow agricultural robots to capture a comprehensive view of their surroundings in real time, which is necessary for detecting objects and obstacles from all directions, facilitating safe and efficient navigation through fields and orchards. However, this high-resolution and accuracy of LiDAR-generated maps can be computationally intensive, demanding robust hardware and processing power [67]. LiDAR sensor limitations, influenced by factors such as weather conditions (i.e., heavy rain or fog) and occlusions, can impact the accuracy and reliability of SLAM algorithms and, consequently, the navigation of agricultural mobile robots. Additionally, in environments with limited visual features or complex and chaotic spaces, SLAM may face challenges [68]. The cost of high-quality LiDAR sensors, which typically range from one thousand to several thousand dollars, can be a significant factor contributing to the overall production costs of the robot [68], making the robot less affordable for some farmers. A list of LiDAR sensors and their specifications that can be used in agricultural mobile robots is presented in Table 1.5. Here the Field of View (FoV) indicates the extent of the LiDAR's scanning range, which is essential for assessing its suitability for specific navigation tasks within agricultural settings. The minimum and maximum distance range is also an important factor to be considered for obstacle detection and avoidance.

TABLE 1.5

Commonly Used LiDAR Sensors Used for Autonomous Navigation of Agricultural Mobile Robots

LiDAR Sensor: Name, Model	Field of View (degree °)	Range (m), (min,max)	Precision (cm)	Weight (g)	Resolution (degree °)	Scan Rate (points/s)	Wavelength (nm)	Power (W)	Interface	Price Range (USD)
Velodyne VLP-16, Puck	360 H	0.15–100	2	830	0.2°–2°	600,000	905	8–16	Ethernet	3,000–4,000
Velodyne HDL-32E, HDL-32	360 H	1–100	2	2,300	0.08°	700,000	905	38	Ethernet	20,000–30,000
Velodyne Alpha Prime, Alpha Prime	360 H	0.2–200	2	4,900	0.025°	1.3 million	905	60–150	Ethernet	Price on request
SICK LMS511, LMS511	270 H	0.05–80	2	1,000	0.167°	1000	905	7.5	Ethernet/RS-232	3,000–5,000
SICK LMS400, LMS400	360 H	0.05–50	2	1,040	0.25°	400	905	7.5	Ethernet/RS-232	1,000–3,000
Hokuyo UST-10LX, UST-10LX	270 H	0.06–10	1	370	0.25°	40,000	905	5.6	Ethernet/USB	2,000–3,000
Hokuyo URG-04LX, URG-04LX	240 H	0.02–4	3	160	1°	10,000	905	5.5	USB	1,000–2,000
Livox Mid-40, Mid-40	38.4 V, 360 H	0.1–260	2	710	0.1°	100,000	905	25	Ethernet	1,000–2,000
Ouster OS-1, OS-1	64 V, 360 H	0.1–120	2	1,100	0.033°	1.28 million	850	20	Ethernet	5,000–6,000
Quanergy M8, M8	360° H	0.1–150	2	1,300	0.033°	500,000	905	30	Ethernet	Price on request

1.3.3 SENSORS FOR OBSTACLE DETECTION

Detection of obstacles in the field, such as random rocks and bushes, animals, humans, and trees, requires a wide range of sensors, including vision-based sensors [35] (such as visible, infrared, thermal imaging, and hyperspectral imaging), as well as distance detection sensors (such as LiDAR, ultrasonic sensors, and laser range finders). The outputs of these sensors are usually fused [69] to generate a comprehensive and accurate representation of the environment and for planning a safe path [70]. Cameras and imaging technologies encompass a diverse array of tools for capturing visual and non-visible information. A list of different types of vision sensors with a short description and examples for each case is presented in Table 1.6. For example, RGB-D cameras [71], combining RGB capabilities with depth sensing, provide detailed color images and are fundamental for color recognition and object detection, enabling 3D modeling and depth perception. This method uses two or more cameras to create a 3D image of the robot's environment, providing the robot with information about distance to obstacles. Stereovision [72, 73] is particularly useful in agricultural environments where there is a lot of visual complexity, as it provides a more detailed view of the environment than single-camera systems can. Thermal cameras [38] detect heat radiation rather than visible light and are used to capture thermal images that display temperature variations in agricultural fields, which are used for night vision navigation, detecting animals and humans. Infrared cameras [74] capture images in the infrared spectrum, beyond the range of human vision, and are also used for detecting living organisms as they can detect heat signatures and are unaffected by low-light conditions.

Multi-spectral cameras [83] capture images in multiple wavelength bands across the electromagnetic spectrum, beyond visible light. These cameras are used for specialized obstacle detection applications, such as specific plants or objects where different spectral bands reveal specific information about objects or terrain being observed. Hyperspectral cameras [79, 84, 85] capture images in many narrow, contiguous spectral bands, providing detailed spectral information for each pixel in an image. This technology is of great interest in precise identification of specific plants and objects in the field. A 360-degree camera [86] captures a panoramic view of the robot surrounding in all directions simultaneously and is often used on agricultural mobile robots for 3D photography and virtual reality, providing a complete view of the field for navigation and collision avoidance [81, 87]. A Pan-Tilt-Zoom (PTZ) camera [81, 88] is equipped with motorized mechanisms that enable it to pan (rotate horizontally), tilt (rotate vertically), and zoom in and out. Although originally designed for surveillance and remote monitoring, PTZ cameras offer unique advantages when integrated into agricultural robotic systems by providing flexibility for adjusting the robot's field of view and focus, allowing it to effectively scan the environment for obstacles such as crops, terrain irregularities, or unexpected objects. A fish-eye camera [89, 90] benefits from a specialized wide-angle lens to capture an exceptionally wide field of view (often exceeding 180 degrees) and can provide a comprehensive scan of the field, aiding the robot in detecting obstacles, terrain irregularities, and potential hazards across a broad perspective.

TABLE 1.6

Type of Vision-Based Perception Used in Agricultural Mobile Robots for Detection of Obstacles, Humans, Animals, and Plants

Type of Vision Sensor	Description	Example Sensor
RGB [71]	Standard color camera capturing red, green, and blue channels for visual perception and object detection	—FLIR Chameleon3 —Basler Ace Series —Allied Vision Mako Series
RGB-D [75]	Combines RGB imaging with depth sensing, allowing for 3D perception and depth mapping	—Microsoft Azure Kinect —Intel RealSense D400 Series —Orbbec Astra Series
Stereo Camera [76]	Consists of two or more lenses to capture depth information, enabling stereo vision and 3D reconstruction	—ZED by Stereolabs —Intel RealSense T265 —Bumblebee by FLIR
Thermal Camera [38]	Captures heat signatures and temperature variations, often used for monitoring and detecting anomalies in agriculture	—FLIR Lepton Series —FLIR A Series —Seek Thermal Compact Series
Infrared (IR) [77]	Captures images in the infrared spectrum, useful for tasks such as night vision and temperature monitoring	—FLIR Ex Series —FLIR T Series —Axis Q19 Series
Multi-Spectral [78]	Captures images in multiple spectral bands, facilitating detailed analysis of vegetation health and crop characteristics	—MicaSense RedEdge —MX—Parrot Sequoia+ —Tetracam Mini-MCA6
Hyperspectral [79]	Provides high-resolution data in numerous narrow, contiguous spectral bands for advanced remote sensing applications	—Headwall Photonics Nano —Hyperspec —Specim AISA Eagle —NEO Hyspex VNIR-1600
360-Degree [80]	Captures a full 360-degree view of the surroundings, suitable for monitoring and navigation in all directions	—Insta360 ONE X2 —GoPro Max —Ricoh Theta Z1
Pan-Tilt-Zoom (PTZ) [81]	Can be remotely controlled to pan, tilt, and zoom, offering flexible surveillance and monitoring capabilities	—Axis Communications PTZ Series —Hikvision DS-2DF Series —Sony SRG Series
Fish-Eye [82]	Utilizes a wide-angle lens to capture a hemispherical view of the surroundings, often used for panoramic imagery	—GoPro Fusion —Insta360 EVO —Ricoh Theta V

Some of the most widely used RGB and RGB-D cameras for obstacle detection are listed in Table 1.7, along with a summary of their key features and price ranges. One of the most popular sensors is the Intel RealSense D435 stereo camera [91], which combines depth perception with RGB imaging, enabling accurate obstacle detection and 3D mapping. Its compact form factor makes it a suitable choice for integration into agricultural robotic systems. Similarly, the D435i includes an IMU

TABLE 1.7
Commonly Used RGB and RGB-D Cameras for Obstacle Detection

Camera Model	Key Features	Resolution	Field of View (Degree °)	Frame Rate (FPS)	Price Range
Intel RealSense D435 [91]	RGB-D with stereo cameras. Ideal for SLAM and obstacle detection.	1280x720 RGB, 1280x720 Depth	86.2 by 57.4	90	$150–$200
Intel D435i Stereo Camera [93]	Combines RGB-D and IMU data for improved navigation, suitable for lightweight mobile robots	1920x1080 RGB, 1280x720 Depth	85.2 by 58	90	$250–$350
Intel RealSense D455 [94]	RGB-D high-resolution, suitable for navigation and object recognition	2560x720 RGB, 2560x720 Depth	86.2 by 57.4	90	
Intel Realsense T265 [95]	Visual SLAM camera with tracking and mapping capabilities, compact and lightweight	N/A	163 by 106	30	$199–$249
Stereolabs ZED Mini Stereo Camera [96]	Lightweight and small form factor stereo camera, offers depth perception for obstacle avoidance	2560x720 RGB, 256x720 Depth	110 by 124	100	$199–$449
Stereolabs ZED 2 Stereo Camera [97]	RGB-D high-resolution stereo vision, suitable robotics and mapping	4416x1242 RGB, 1280x720 Depth	90 by 60	30	$449–$899
Microsoft Azure Kinect DK [98]	High-definition RGB and depth sensing, suitable for 3D mapping, gesture recognition, and robotics	3840x2160 RGB, 640x576 Depth	75 by 65	30	$399–$499
JeVois-A33 Smart Camera [99]	Compact and programmable smart camera, suitable for a wide range of vision-based navigation	640x480 RGB	Varies	30	
OpenCV OAK-D Camera [92]	Open-source RGB-D camera with AI capabilities, ideal for object detection, and mapping	1280x720 RGB, 1280x720 Depth	Varies	30	$199–$299
Luxonis Movidius DepthAI OAK-1 [100]	AI-powered RGB-D camera with depth and object recognition capabilities. Compact and customizable.	1280x720 RGB, 1280x720 Depth	Varies	30	$199–$249

for enhanced positioning and obstacle detection capabilities, making it ideal for operating in challenging terrains. The RealSense D455 offers an extended range for depth perception, allowing robots to detect obstacles at greater distances, a valuable feature for open-field agricultural applications. The current trend is moving toward cameras with AI features, such as the OpenCV OAK-D [92] and the Depth AI OAK-1 cameras, which leverage AI capabilities for complex obstacle detection and real-time avoidance strategies.

The safety system of an agricultural mobile robot can greatly benefit from custom-built human detection sensors or the integration of ready-to-use cameras equipped with built-in human detection [101] and night vision [102] features, such as the ones listed in Table 1.8. Besides safety features, these cameras can also enhance the operational capabilities during low-light conditions or nighttime scenarios. An example of the camera models with built-in human detection feature is the Google Nest Cam IQ [103], which is known for its high-resolution imaging and intelligent human detection capabilities. It provides real-time alerts when a human is detected in the robot's vicinity, even during night operation. The Reolink Argus PT [104] is a versatile camera equipped with pan-and-tilt functionality [105] for human detection and night vision. Arlo Ultra 2 [104] offers 4K video quality, advanced human detection, and enhanced night vision capabilities to ensure precision in identifying human activity and provides clear visuals, even in complete darkness. Additionally, a range of Lorex Security Cameras, the Ring Stick Up Cam, Wyze Cam V3, Blink Outdoor Camera, Amazon Ring Indoor Cam, EufyCam 2C, and TP-Link Kasa Spot cameras can contribute to the safety and security of agricultural mobile robots with their reliable human detection and night vision capabilities. Figure 1.4 shows examples of human detection using RGB-D sensors with corresponding depth maps and reference maps for an autonomous lawn mowing operation [71].

Some agricultural mobile robots may benefit from one or multiple arrays of distance detection sensors, such as the one shown in Figure 1.6. A list of typical distance detection sensors that can be used for obstacle detection under different light and weather conditions is given in Table 1.9. These sensors employ a variety of distance detection methods, including Ultrasonic, ToF and IR [44, 106], LiDAR, and Stereo Vision, each chosen for its unique strengths in specific scenarios. For example, the low-cost and versatile Parallax PING))) and HC-SR04 ultrasonic sensors [107] are two distance detection sensors widely used for their simplicity and reliability. Operating on the principle of ultrasonic echolocation, these sensors emit high-frequency sound waves, which then bounce off objects in their vicinity. By precisely measuring the time it takes for these sound waves to return to the sensor, these sensors calculate the distance to the object with impressive accuracy.

The Sharp GP2Y0A21YK0F [108], Adafruit VL53L0X, and Adafruit VL53L1X are other popular IR sensors providing laser-like accuracy in distance measurement. In specific, the Sharp GP2Y0A21YK0F is widely used in multi-array setups such as the one shown in Figure 1.7 due to its precision and robust performance in a wide range of field conditions with variations in lighting and temperature. They provide

TABLE 1.8

Ready-to-Use Cameras with Built-In Human Detection and Night Vision Features for Safety of Agricultural Mobile Robots

Camera Model	Field of View (FoV)	Resolution	Interface(s)	Weight (g)	IP Rating	Price Range
Google Nest Cam IQ	130° H, 75° V	1080p	Wi-Fi, Ethernet	320 g	IP54	$129–$299
Reolink Argus PT	105° H, 57° V	1080p	Wi-Fi, Ethernet	408 g	IP65	$99–$149
Arlo Ultra 2	180° H, 125° V	4K UHD	Wi-Fi	454 g	IP65	$249–$499
Lorex Security Cameras	Varies by model	Varies by model	Ethernet	Varies by model	Varies	$79–$299
Ring Stick Up Cam	150° H, 85° V	1080p	Wi-Fi, Ethernet	370 g	IPX5	$79–$99
Wyze Cam V3	110° H, 59° V	1080p	Wi-Fi	90 g	IP65	$23–$32
Blink Outdoor Camera	110° H, 55° V	1080p	Wi-Fi	77 g	IP65	$79–$99
Amazon Ring Indoor Cam	115° H, 60° V	1080p	Wi-Fi, Ethernet	68 g	IPX5	$59–$59
EufyCam 2C	140° H, 65° V	1080p	Wi-Fi	490 g	IP67	$99–$149
TP-Link Kasa Spot	130° H, 80° V	1080p	Wi-Fi	62 g	IP64	$39–$45

FIGURE 1.5 Example of human detection using RGB-D sensor and reference maps for an autonomous lawn mowing robot [71].

FIGURE 1.6 Multi-channel IR distance detection sensors used on electrical tractors for collision avoidance in orchards.

analog output voltage that is directly proportional to the measured distance and can be easily interfaced with microcontrollers. The Benewake TFmini and TFmini plus, Livox Mid-40 LiDAR, Hokuyo URG-04LX, Sick LMS511, RPLidar A2M8, Velodyne VLP-16, and Intel RealSense D435 are among the LiDAR and depth-sensing obstacle detection solutions that can also be used for mapping and creating detailed 2D and 3D representations of the surroundings. The Benewake TFmini and TFmini Plus are two affordable LiDAR-based distance detection solutions that emit laser beams and measure the time it takes for the beams to bounce back after hitting objects in their surroundings, providing highly accurate distance measurements. Both sensors are compact and suitable for integration into agricultural robotic systems where space constraints should be considered. They offer fast response and reliable measurements for close-range obstacle detection (below 6 meters) and can operate effectively in various environmental conditions, including bright sunlight and low-light scenarios.

TABLE 1.9

Distance Detection Sensors Used in Obstacle Detection, Collision Avoidance, and Mapping for Agricultural Mobile Robots

Sensor Model	Method	Detection Range (Approx.)	Accuracy (Approx.)	Communication Interface	Price Range (Approx.)
Parallax PING)))	Ultrasonic, ToF	3 m	1–2 mm	analog and digital	$5–$10
HC-SR04	Ultrasonic, ToF	2 cm–400 cm	3 mm	GPIO, PWM	$2–$5
MaxBotix (short, medium, long) Range Finders	Ultrasonic, ToF	20 cm–30 m	3 cm (±3%)	Analog, UART, PWM	$20–$300
Sharp GP2Y0A21YK0F	IR, analog triangulation	10 cm–80 cm	±10%	Analog	$10–$20
Adafruit VL53L0X	IR, ToF	Up to 2 meters	±3%	I2C	$5–$15
Adafruit VL53L1X	IR, ToF	Up to 4 meters	±1.5%	I2C	$10–$20
Benewake TFmini and TFmini plus	LiDAR	30 cm–12 meters	±5%	UART, I2C	$30–$50
Livox Mid-40 LiDAR	LiDAR	260 m	±2%	Ethernet	$600
Hokuyo URG-04LX	LiDAR	4.0 meters	±30 mm	USB, RS-232	$900–$1,500
Sick LMS511	LiDAR	Up to 80 meters	±50 mm	Ethernet, CAN	$4,000–$6,000
RPLidar A2M8	LiDAR	6 meters	±1%	UART, USB, ROS	$200–$400
Velodyne VLP-16	LiDAR	Up to 100 meters	-	Ethernet, CAN	$5,000–$8,000
Intel RealSense D435	Stereo Vision	Up to 10 meters	-	USB, ROS	$200–$400

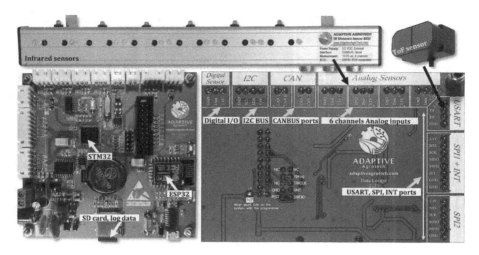

FIGURE 1.7 An obstacle detection sensing system with multi-array infrared sensors and ToF laser sensors with CANBUS communication used in the collision avoidance of an electrical tractor [106].

1.4 ALGORITHMS FOR AUTONOMOUS NAVIGATION AND COLLISION AVOIDANCE

The research and development on mobile robots to perceive the surroundings and navigate intelligently and efficiently in complex environments has led to the development of two fundamental approaches, heuristic [109–111] and optimal navigation [112–115], as shown schematically in Figure 1.8. In the heuristic approach, autonomy is accomplished through a set of practical rules or behaviors, which do not necessarily guarantee an optimal result, but are good enough to achieve an immediate goal. In this approach, complete information about the environment is not required because the autonomous navigation relies on practical rules of thumb or "heuristics" to make decisions. Instead of seeking a mathematically guaranteed optimal solution, heuristic algorithms prioritize quick and practical decision-making based on experience or domain-specific knowledge. Heuristic methods are often computationally efficient and can make decisions in real time because they prioritize speed and responsiveness, which is crucial in dynamic environments. In addition, heuristic algorithms tend to be relatively straightforward and easy to implement. They rely on simple rules or strategies that do not require complex optimization or extensive computation, hence may not always yield the absolute best solution but aim to find a good enough solution that meets specific criteria or objectives. Heuristic approaches include methods such as Potential Fields [116, 117], Bug Algorithms [118–120], and behavior-based navigation [121–123], which use rules based on proximity to obstacles and predefined goals to guide the robot's motion. These algorithms can adapt to changing situations by adjusting their rules or heuristics based on new sensory information or environmental conditions.

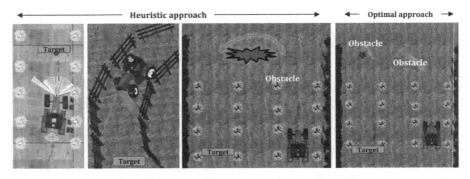

FIGURE 1.8   Representation of heuristic and optimal approach for collision avoidance of an autonomous tractor.

In the optimal approach, planning is achieved through mathematical optimization to rigorously search for the best path or trajectory that minimizes or maximizes an objective function, such as travel time, travel distance, energy consumption, or safety. This requires more environment information because it seeks to find the best solution based on a well-defined objective function. The outcome guarantees that the robot selects the most efficient and collision-free path [124, 125] while taking into account some specific constraints. However, these algorithms can be computationally intensive, especially in high-dimensional spaces or complex environments, and hence may not be suitable for real-time applications without significant simplifications. Many optimal approaches focus on global path planning, considering the entire environment and goal location when generating a solution. Algorithms like A* search [126], Dijkstra's algorithm [127, 128], Rapidly-exploring Random Trees (RRT) [129–131], and model predictive control (MPC) [132] fall under the optimal approach. Table 1.10 summarizes a list of algorithms that are commonly used in localization, mapping, path planning, path tracking, sensor fusion, and collision avoidance of agricultural mobile robots. In practice, the choice between heuristic and optimal approaches in autonomous navigation depends on the specific requirements of the robotic system, the computational resources available, the complexity of the environment, and the desired trade-off between speed and solution quality. Although heuristic methods offer practical solutions for many real-world scenarios, optimal methods are valuable when precision and guarantees are priorities, even if they come at the cost of increased computational complexity.

1.4.1   LOCALIZATIONS AND MAPPING

Localization and mapping algorithms enable the robot to determine its position in an environment and create a map of that environment. For example, Kalman Filter [158, 159] is used as a key component of many localization systems for estimating the state of the robot when measurements are uncertain and subject to noise. It is an optimal recursive algorithm that models the system's state by incorporating sensor measurements and system dynamics to minimize the mean square error of the estimated state. Extended Kalman Filter [134, 135] is also a recursive algorithm that combines sensor

Sensors, Algorithms, and Software for Autonomous Navigation

TABLE 1.10

List of Algorithms Commonly Used in Localization, Mapping, Path Planning, Path Tracking, Sensor Fusion, and Collision Avoidance of Agricultural Mobile Robots

Algorithm	Heuristic/ Optimal	Main Sensors Required	Limitations
Localization			
Kalman Filter [133]	Optimal	Odometry, GPS, IMU, Encoders	Assumes linear system dynamics and Gaussian noise. May not perform well with nonlinear or highly dynamic systems.
EKF [134, 135]	Heuristic	Odometry, GPS, IMU	Sensitive to sensor noise and drift. Limited accuracy in environments with significant signal loss or multipath interference.
Particle Filter [133]	Heuristic	Odometry, GPS, IMU	Computational complexity increases with the number of particles. Convergence can be slow in environments with multimodal distributions.
Mapping			
Grid-based Occupancy, [136–138] Grid Mapping [139, 140]	Optimal	Lidar, RGB-D camera, Sonar, Laser Range Finder	Limited to static environments. Doesn't handle dynamic obstacles well. Resolution may impact accuracy.
SLAM [141]	Optimal	Lidar, RGB-D camera, Sonar, Laser Range Finder	Complexity increases with map size and feature-rich environments. Mapping errors can accumulate over time.
Graph-Based SLAM [142] (e.g., G2O, g2o)	Optimal	Lidar, RGB-D camera, Sonar, Laser Range Finder	Requires reliable data association. Computational load can be high for large-scale environments.
Path Planning			
Dijkstra's [127, 128]	Optimal	Map (occupancy grid), GPS	Can be computationally expensive for large graphs. Doesn't consider dynamic obstacles. May not always find a feasible path.
A* [126]	Optimal	Map (occupancy grid), GPS	The quality of the heuristic impacts performance. Suboptimal solutions possible if the heuristic is not admissible.
RRT [131]	Heuristic	Lidar, RGB-D camera, Sonar, Laser Range Finder	Lack of optimality. Path quality depends on exploration strategy. Not ideal for deterministic environments.

(Continued)

TABLE 1.10 *(Continued)*
List of Algorithms Commonly Used in Localization, Mapping, Path Planning, Path Tracking, Sensor Fusion, and Collision Avoidance of Agricultural Mobile Robots

Algorithm	Heuristic/ Optimal	Main Sensors Required	Limitations
Path tracking			
Pure Pursuit [143]	Heuristic	Odometry, GPS, Lidar, Encoders	Doesn't perform global path planning. Limited to following predefined paths. May not handle dynamic obstacles well.
Stanley Controller [144, 145]	Heuristic	GPS, IMU, Odometry, Map	Sensitive to sensor noise and drift. Requires a map of the path for reference.
LQR [146]	Optimal	Odometry, IMU, Model	Requires an accurate model of the robot's dynamics. May not handle nonlinear systems well without adaptations.
MPC [132]	Optimal	Encoders, IMU, Odometry, Lidar, RGB-D camera	Computational complexity. Real-time requirements can be challenging. Sensitive to model inaccuracies.
PID Control [147]	Heuristic	Encoders, IMU, Odometry	Performance may degrade in nonlinear or dynamic systems. Requires careful tuning. Sensitivity to model inaccuracies.
Line-of-Sight (LoS) [148, 149]	Heuristic	Odometry, GPS, Lidar	Sensitive to sensor noise and limited visibility. May require frequent path updates.
Sensor Fusion			
Unscented Kalman Filters (UKF) [150]	Optimal	IMU, GPS, Lidar, Radar, Camera	Requires selecting appropriate sigma points. Complexity increases with higher-dimensional state spaces.
Complementary Filter [151, 152]	Heuristic	Accelerometer, Gyroscope	Prone to drift over time and may require periodic calibration. Limited accuracy for long-term applications.
Information Fusion [153]	Optimal	Lidar, Radar, Camera, IMU, GPS	Requires maintaining and propagating information matrices. Complex to implement and tune.
Bayesian Filters [154]	Optimal	Lidar, Radar, Camera, IMU, GPS	Computational complexity can be high, especially in high-dimensional state spaces. Sensitivity to mwwodel inaccuracies.
Collision Avoidance			
Potential Fields [116]	Heuristic	Lidar, Sonar, Infrared Sensors	May get stuck in local minima. Tuning the field parameters can be challenging. Sensitivity to obstacle shape and density.

TABLE 1.10 *(Continued)*
List of Algorithms Commonly Used in Localization, Mapping, Path Planning, Path Tracking, Sensor Fusion, and Collision Avoidance of Agricultural Mobile Robots

Algorithm	Heuristic/ Optimal	Main Sensors Required	Limitations
Braitenberg [155]	Heuristic	proximity sensors such as infrared or ultrasonic, Lidar	Lack of global planning, no memory of past actions, sensitivity to sensor noise, limited adaptability, not suitable for all environments.
Dynamic Window Approach [156]	Heuristic	Lidar, Sonar, Infrared Sensors	Reactive and may not always plan globally. Performance depends on the choice of dynamic window parameters.
Follow the Gap [157]	Heuristic	Lidar, Sonar, Infrared Sensors	May not work well in highly cluttered or complex environments. Sensitivity to gap detection algorithm.

measurements (e.g., odometry, GPS, IMU) with a motion model to estimate the robot's position and orientation. It maintains a probability distribution over the robot's pose and updates it with new sensor data. Particle Filter [133] or Monte Carlo Localization (MCL) [160] utilizes a set of particles to estimate and represent possible robot poses and updates their weights based on sensor data. It provides a probabilistic estimate of the robot's pose, making it suitable for environments with uncertainty. MCL is a local localization technique in the sense that it estimates the robot's pose within a known map of its immediate surroundings. It does not inherently provide a global pose estimate over a large-scale map. However, by using MCL iteratively and incorporating information from multiple places within the environment, a robot can gradually build a more global estimate of its pose. To achieve global localization or navigation, MCL can be combined with global path planning algorithms (such as Dijkstra's algorithm or A*) to navigate the robot through a map while keeping track of its pose using MCL. This combination allows the robot to navigate globally while continuously updating its local pose estimate to avoid obstacles and maintain accuracy.

Grid-based Occupancy [161] and Grid Mapping [139, 140] create environment maps by dividing the environment into a grid of cells and assigning occupancy probabilities to each cell based on sensor measurements. It produces an occupancy grid map, which represents the likelihood of each cell being occupied or free. SLAM combines estimating the robot's pose and the map of the environment concurrently using sensor data, and is particularly valuable for exploring and mapping unknown environments in real time. Graph-Based SLAM [142] such as G2O and g2o builds a map and refines the robot's pose estimation by modeling the environment as a graph, where nodes correspond to robot poses (based on sensor measurements) and map features, minimizing errors and improving accuracy over time.

1.4.2   PATH PLANNING

Global path planning (GPP) [162] is concerned with finding a path from the robot's current position to its desired destination (goal) over a large-scale map of the environment, including static obstacles. The goal is typically specified in a global coordinate frame. Algorithms such as Dijkstra's [163], A* [164], Rapidly-exploring Random Trees (RRT) [130], Voronoi diagrams [165], and various heuristic-based methods are frequently used. Dijkstra's is a graph search algorithm that computes the shortest path between two points in a graph (representing the environment) while considering edge costs (e.g., distances) [25]. It is suitable for scenarios where finding the global optimal path is essential. A* is an extension of Dijkstra's algorithm that uses a heuristic function to prioritize nodes likely to lead to the goal [166]. It is more efficient than Dijkstra's algorithm for path planning and is widely used in robotics for finding optimal paths. RRT grows a tree of random nodes in the configuration space, aiming to find a path to the goal while efficiently exploring and planning in an unknown environment [167]. It is particularly useful in scenarios with unknown or complex environments.

It should be noted that GPP is usually performed less frequently because it involves a more computationally intensive search for an optimal or near-optimal path [162]. Local Path Planning (LPP) [168], also known as local motion planning [169], focuses on generating immediate, short-term paths to navigate around obstacles and maintain safe distances. Techniques such as Dynamic Window Approach [156, 157, 170, 171] and trajectory generation are used for this purpose. LPP's goal is to generate a safe and feasible trajectory for the robot to follow within a short time horizon, considering dynamic obstacles, sensor feedback, and real-time constraints. It operates at a lower level and takes into account the robot's current state and sensor data, and is performed at a high frequency, typically on the order of milliseconds, to respond rapidly to changing conditions and obstacles. The planner continuously updates the robot's path based on real-time sensor information to avoid collisions and adapt to changes in the environment. Local path planners often use techniques such as obstacle avoidance algorithms, velocity control, and reactive planning [172]. It should be noted that many robotic systems employ a combination of global and local path planning strategies to achieve a balance between long-term and short-term navigation goals. Dynamic path planning [173] is another trend in the collision avoidance of agricultural mobile robots, allowing the robot to generate safe and efficient navigation paths in real time, based on the current state of the environment [162]. It can be used to avoid obstacles, optimize task performance, and respond to changes in the environment, such as sudden obstacles or changes in terrain.

1.4.3   PATH TRACKING

Path tracking [174] is the process of controlling a robot to follow a predefined path as closely as possible. It is typically executed in real time, continuously adjusting the robot's control inputs (e.g., steering angle and velocity) to stay on the path as it encounters variations and disturbances in the environment. The path is usually represented at a relatively low level of detail, often as a sequence of waypoints or reference points in space connected by straight-line segments or smooth curves. It relies heavily on feedback control techniques to correct deviations from the path. Common methods include Pure Pursuit [175], PID Controllers [176], and Model Predictive Control (MPC) [177].

The Pure Pursuit control algorithm works by selecting a target point ahead of the robot's current position and calculating control commands, such as steering angles, to navigate the robot along the desired path. Stanley Controller [144] utilizes feedback control based on the cross-track error [178] (lateral distance between the robot and desired path) to calculate the steering angle. LQR [179] is an optimal control algorithm that can be adapted for path tracking by minimizing a cost function to optimize control inputs such as steering and acceleration. Model Predictive Control [180] predicts the robot's future behavior over a finite time horizon and computes control inputs that optimize a predefined cost function, considering constraints. MPC is suitable for systems with complex dynamics and can handle both local and global planning objectives. Line-of-Sight (LoS) [181] guides a mobile robot along a desired path by continuously calculating the desired heading angle based on the LoS to a target point. It adjusts the robot's steering angle accordingly to keep it aligned with the path. As the robot moves, it updates its position and makes real-time corrections to maintain precision and stay on track.

1.4.4 TRAJECTORY PLANNING

Trajectory planning [182] involves generating a set of continuous and feasible motion commands (trajectories) that the robot can follow to reach its destination while considering various constraints, including kinematics, dynamics, and obstacle avoidance. Trajectory planning considers the entire motion from the initial state to the final state, taking into account not only the path but also the robot's speed, acceleration, and dynamics. Trajectory planning typically operates at a higher level of detail, specifying the robot's motion over time, including velocity profiles, accelerations, and steering commands. Trajectory planning can be performed offline or online. Offline trajectory planning [183] generates a set of motion commands in advance, whereas online planning [184] adjusts the trajectory in real time based on the current environment. Trajectory planning considers the robot's dynamic constraints, ensuring that the generated trajectory is feasible in terms of the robot's physical capabilities.

1.4.5 SENSOR FUSION ALGORITHMS

Sensor fusion algorithms [62] combine data from various sensors such as LiDAR, cameras, GPS, IMUs, and proximity sensors to create a more comprehensive and accurate perception of the environment for the robots. Commonly used algorithms include Kalman Filters (KF) [185], extended KF [186], Unscented KF (UKF), Particle Filters [133], Complementary Filters [187], Visual-Inertial Odometry (VIO) [188], and LiDAR-IMU fusion [189]. A list of some of the agricultural mobile robots that benefit from sensor fusion is provided in Table 1.11. Unscented Kalman Filter, an alternative to the EKF, is suitable for sensor fusion in nonlinear systems. It approximates the probability distribution of the state through a set of sample points, known as sigma points, and can be used for fusing data from sensors such as IMUs, GPS, and LiDAR for accurate state estimation. Complementary Filter combines

TABLE 1.11
List of Agricultural Mobile Robots That Benefit from Sensor Fusion

Robot Name	Fused Sensor Data
AgBot [190]	Employs multi-sensor fusion, including Lidar, cameras, and GPS, for crop monitoring and weeding
Ladybird [191]	Utilizes Visual-Inertial Odometry (VIO) to navigate autonomously in agricultural fields, combining camera and IMU data for accurate localization
Thorvald [17]	Employs GPS-IMU sensor fusion to navigate and perform tasks such as plant phenotyping and mapping in outdoor environments
Prospero [192]	Uses a combination of Lidar and camera data for navigation, crop monitoring, and obstacle detection in vineyards and orchards
BoniRob [193]	Integrates radar and GPS data to navigate through fields, enabling autonomous navigation and precision agriculture tasks
Blue River Technology [194]	The See & Spray system of Blue River Technology uses multispectral camera data fusion for precision spraying of herbicides in row crops, reducing chemical usage.

accelerometer and gyroscope data to estimate the orientation of the robot by using low-pass and high-pass filters to reduce noise. Information Filter is an alternative to the Kalman Filter that works with information matrices instead of state estimates, and is suitable for sensor fusion in systems with multiple sensors, such as LiDAR, radar, and cameras. It can provide accurate estimations for both linear and nonlinear systems. Bayesian Filters such as the Bayesian Recursive Estimation (Bayes' filter) can fuse data from various sensors using Bayesian inference. These filters are versatile and can be applied to various sensor combinations, including LiDAR, radar, cameras, and IMUs, for state estimation and object tracking.

1.4.6 COLLISION AVOIDANCE ALGORITHMS

Collision avoidance algorithms work based on two approaches, reactive control [195] and deliberative planning [196]. In reactive control, the robot collects information about surroundings using built-in obstacle-detecting sensors, performs reactive maneuvers based on this information, and responds to its environment in a direct and immediate way. This type of approach enables quick responses to unpredicted changes in environment because actions are not pre-planned; instead, the system reacts and makes decisions on the go, generally without considering the future implications of those decisions. Reactive control algorithms are often based on simple, rule-based decision mechanisms, and are capable of operating in real-time environments due to low computational requirements. Nevertheless, the reactive control does not involve any forward-looking planning or consideration of future states, and therefore control may contribute to a local minimum. Examples of the reactive control method include the potential field [116], Braitenberg [197], Dynamic Window Approach [170], and Follow the Gap algorithms [198]. The Potential Fields algorithm attracts the robot to the goal and repels it from obstacles using potential fields. It adjusts the robot's velocity based on the attractive and repulsive forces to help navigate while avoiding collisions. Braitenberg relies on simple rules and reactive behaviors based on sensor inputs, making quick and local decisions to navigate and avoid obstacles in real time. This heuristic nature allows it to be computationally efficient and suitable for rapid response in dynamic environments; however, it may lead the robot into situations where it becomes stuck in complex environments. The Dynamic Window Approach considers the robot's kinematics and dynamics to compute a feasible velocity command while avoiding obstacles within a short time horizon. It evaluates the robot's future states to ensure safe and efficient navigation. Follow the Gap is a collision avoidance navigation technique that identifies and follows "gaps," or open areas between obstacles, to traverse through narrow passages. It is valuable for robots operating in cluttered environments or agricultural fields.

In contrast, in deliberative planning the robot collects surroundings using obstacle sensors and generates a route map for decision making by considering and evaluating possible future states and actions. After that, an optimal path [115] with a collision-free route [199] is determined using the initial position as a reference point and following that route plan to execute. This approach relies on creating a plan in advance, typically considering a sequence of actions that will lead to achieving a certain goal. Deliberative planning algorithms consider future states and consequences of actions, are often computationally intensive due to the need to explore potential future actions and states, are goal-oriented (aims to achieve specific objectives or goals), and are less responsive (slower in making decisions due to thoroughness in planning). Many modern agricultural mobile robots benefit from a hybrid strategy [200], combining aspects of both reactive and deliberative planning depending on the needs of the environment. In a hybrid model, a system may utilize deliberative planning to formulate overall strategies and goals, while employing reactive control to manage immediate and unforeseen issues. This approach seeks to balance the fast, real-time responsiveness of reactive control with the foresight and goal-oriented nature of deliberative planning.

1.4.7 MACHINE LEARNING

Machine learning (ML) [201] algorithms have emerged as a powerful tool to enhance the capabilities of mobile robots, by training them to recognize and avoid obstacles, even in unpredictable environments. They are also used to improve the accuracy of sensors, such as cameras and LiDARs, by compensating for environmental factors such as lighting and weather conditions. By training on historical navigation data, mobile robots can learn optimal paths through fields, considering factors such as terrain roughness, soil conditions, and existing obstacles. ML models such as deep neural networks and convolutional neural networks (CNNs) [202] have been used to process images captured by robot-mounted cameras to distinguish between crops and obstacle, beside recurrent neural networks (RNNs) that are used to handle sequential data, such as LiDAR scans or GPS measurements. These models can also forecast future sensor data, allowing the robot to anticipate and react to changes in the environment proactively. Recurrent networks, or variants such as Long Short-Term Memory (LSTM) networks [203], can capture temporal dependencies in sensor data. In addition, ML-based SLAM techniques integrate sensor data to build and update maps while estimating the robot's position [162]. Reinforcement learning (RL) techniques [204] can be applied to train agents on the robots to learn optimal paths or actions through trial and error. RL agents receive rewards or penalties based on their actions and adjust their policies to maximize expected rewards, enabling robots to adapt to changing conditions in real time by navigating around obstacles such as rocks, irrigation systems, or other machinery. Value iteration methods, such as Q-learning [205, 206], can be used for discrete action spaces, while policy gradient methods, such as Proximal Policy Optimization (PPO) [207], work well in continuous action spaces. Object detection models such as Faster R-CNN [208] or YOLO [209], can identify obstacles in RGB images. It should be noted that data quality and quantity are crucial for training accurate ML models, and collecting relevant training data can be time-consuming. Model robustness and adaptability to different agricultural environment conditions are ongoing research areas.

1.5 THE ROBOT OPERATING SYSTEM (ROS) PACKAGES

ROS [210] has become a foundational platform, with a comprehensive ecosystem of packages to facilitate message exchange and the creation of sophisticated and efficient algorithms and strategies for navigating complex environments while avoiding obstacles. ROS is designed with a modular architecture, making it easy to integrate GPS and other sensors, such as LiDARs and IMUs, and provides visualization tools such as *Rviz* [211] for real-time monitoring of sensor data, *costmaps* [212], and robot trajectories [213]. For example, it can be configured to handle temporary GPS signal loss or inaccuracies by using sensor fusion techniques and dead reckoning, ensuring continued navigation in challenging environments. ROS is open-source and has a large and active community of developers that enable access to libraries and save development time and effort. Figure 1.9 shows screenshots from an ROS-based mobile field robot [200] that simultaneously navigates through occluded crop rows and performs various phenotyping tasks, such as measuring plant volume and canopy height using a 2D LiDAR in a nodding configuration [214].

FIGURE 1.9 Using ROS for LiDAR-based autonomous navigation of mobile robots in phenotyping application, showing assembled LiDAR on uneven terrain data visualized within RViZ using the Tilted configuration [200].

GPS-based navigation in ROS involves integrating GPS sensors, acquiring and processing GPS data within ROS nodes, transforming geodetic coordinates into Cartesian coordinates [215], performing sensor fusion for localization and mapping, and using GPS information for global path planning and georeferencing. GPS receivers communicate with the robot's onboard computer through standard communication interfaces such as UART, USB, or Ethernet. Although ROS has drivers available for popular GPS receivers, specific drivers may be required for some other brands to be installed and configured. An ROS node is then created to subscribe to the GPS data stream and convert it into ROS messages, which often include latitude, longitude, altitude, and timestamp information. Since GPS provides geodetic coordinates (latitude, longitude, and altitude) in the Earth's reference frame (WGS84), ROS libraries and functions have been developed (i.e., *geographic_msgs* [216]) to convert these into Cartesian coordinates (x, y, z) in a local coordinate frame, which is more convenient for most robotic applications. ROS also provides packages and libraries that facilitate communication with GPS devices, including *"nmea_navsat_driver"* [217] for NMEA sentence parsing and *"gps_common"* for GPS-related data structures, which are commonly employed in autonomous robots for plant phenotyping and soil sensing [218]. A list of common ROS packages that are used with mobile robots is summarized in Table 1.12.

As discussed earlier, GPS data is prone to noise and multipath errors. To overcome limitations of signal loss as well as improving the localization accuracy and robustness, many agricultural mobile robots use sensor fusion algorithms that combine data from GPS with other sensors. For this purpose, ROS offers the *"robot_localization"* package [219], which leverages Extended [134] or Unscented Kalman Filters [220] to fuse sensor information, such as GPS, IMU, and wheel encoders, assisting accurate localization and obstacle detection. ROS also provides mapping packages such as *"gmapping"* [221] and *"cartographer"* [222] that can be used in conjunction with GPS for SLAM. The core of ROS navigation is the *"move_base"* package [223], which combines global and local planners for path generation and obstacle avoidance. The global planner, often using the *"navfn"* [224] or *"global_planner"* package [225], calculates high-level paths based on global cost maps, while the local planner uses *"base_local_planner"* [226] or *"dwa_local_planner"* [227] to navigate around obstacles using sensor data. The *"move_base"* [228] package accepts GPS coordinates as waypoints and generates trajectories for the robot. These waypoints represent positions along the planned path, enabling the robot to navigate to a goal location using GPS-based guidance. To access additional geospatial information, such as maps, terrain data, and points of interest for enhanced navigation decision-making, ROS can be interfaced with GIS databases and services. For some sensitive robotic farming applications such as autonomous mowing, Geofencing [229] might be implemented using GPS-based boundaries to restrict the robot's movements within specified geographical limits. This is often crucial or even mandatory for ensuring safety and compliance with operational constraints.

Obstacle detection in ROS typically involves segmenting the laser scan data into clusters or regions that correspond to individual obstacles [11]. Clustering algorithms [230], such as Density-Based Spatial Clustering of Applications with Noise (DBSCAN) [231] or Euclidean clustering [232], can be implemented using either

TABLE 1.12

Common ROS Packages Used for Autonomous Navigation and Collision Avoidance of Mobile Robots

ROS Package	Link and Description	Required Sensors
gps_common	https://wiki.ros.org/gps_common Provides essential tools and utilities for parsing, processing, and working with GPS data	GPS
gps_goal	http://wiki.ros.org/gps_goal Converts navigation goals in GPS coordinates to ROS frame coordinates	GPS
Navigation Stack	http://wiki.ros.org/navigation Provides a set of tools for autonomous robot navigation	Lidar, IMU, wheel odometry, GPS, depth camera
move_base	http://wiki.ros.org/move_base Implements a path planning and control framework	Lidar, IMU, wheel odometry, GPS, depth camera
gmapping	http://wiki.ros.org/gmapping Performs laser-based SLAM to create 2D maps of environments	Lidar
hector_slam	http://wiki.ros.org/hector_slam Provides a lightweight 2D lidar-based SLAM solution	Lidar
cartographer	https://github.com/cartographer-project/cartographer Offers 2D and 3D real-time SLAM solutions for accurate mapping	Lidar, IMU, wheel odometry
robot_localization	http://wiki.ros.org/robot_localization Performs sensor fusion for accurate robot pose estimation	IMU, wheel odometry, GPS, magnetometer
amcl	http://wiki.ros.org/amcl Implements adaptive Monte Carlo localization	Laser scan, map
tf	http://wiki.ros.org/tf Manages coordinate frame transformations in a robot system	N/A
costmap_2d	http://wiki.ros.org/costmap_2d Builds costmaps for navigation, considering obstacles	Lidar, depth camera, bump sensors, sonar sensors
obstacle_detector	https://github.com/tysik/obstacle_detector Provides obstacle detection and tracking capabilities from data provided by 2D laser scanners	Lidar, depth camera, ultrasonic sensors, RGB-D camera
teb_local_planner	http://wiki.ros.org/teb_local_planner Offers a trajectory optimization-based local planner	Lidar, IMU, wheel odometry

TABLE 1.12 *(Continued)*
Common ROS Packages Used for Autonomous Navigation and Collision Avoidance of Mobile Robots

ROS Package	Link and Description	Required Sensors
frontier_exploration	http://wiki.ros.org/frontier_exploration Facilitates autonomous exploration of unknown environments	Lidar, IMU, wheel odometry, depth camera
move_base_flex	http://wiki.ros.org/move_base_flex Provides a flexible and modular navigation system	Lidar, IMU, wheel odometry, GPS, depth camera
octomap_mapping	http://wiki.ros.org/octomap_mapping Creates and maintains 3D OctoMap representations	RGB-D camera, Lidar, IMU
rgbdslam	http://wiki.ros.org/rgbdslam A SLAM solution for RGB-D cameras	RGB-D camera
dwa_local_planner	http://wiki.ros.org/dwa_local_planner Implements the Dynamic Window Approach for local planning	Lidar, IMU, wheel odometry
scan_to_cloud_converter	http://wiki.ros.org/scan_to_cloud_converter Converts 2D laser scans to 3D point clouds	Lidar
moveit	http://wiki.ros.org/moveit Provides motion planning and control for mobile robots with manipulators	Arm-specific sensors (e.g., joint encoders, force/torque)

PCL libraries [233] or custom ROS nodes [234]. Raw laser scan data often contains noise and outliers. ROS offers a wide range of sensor drivers and libraries for seamless integration with various types of obstacle detection sensors. For instance, the *"pointcloud_to_laserscan"* package [235] can convert 3D LiDAR point clouds into 2D laser scan data, which is commonly used for obstacle detection. Messages such as *"sensor_msgs/LaserScan"* [236] and *"sensor_msgs/PointCloud2"* [237] are frequently employed to convey sensor data within ROS. These messages summarize information about distance measurements, intensities, and sensor metadata. Deep learning frameworks [238], such as *TensorFlow* [239] and *PyTorch* [240], can be integrated into ROS nodes for advanced obstacle detection tasks. Other libraries, such as the Point Cloud Library *"pcl_ros"* [241], and filters are also available within the *"laser_filters"* package [242] for preprocessing laser data. Common filtering operations include outlier removal, downsampling [243], and smoothing [244]. When clusters of points representing obstacles are identified, object recognition techniques such as shape fitting [245], object classification using machine learning models [246], or pattern recognition [247] can be applied. ROS navigation stack relies on the *"costmap_2d"* package [248] for creating occupancy grid maps and updating costmaps [249] that represent obstacle information and are essential for collision avoidance and path planning. It converts laser scan data into a binary representation where grid cells are labeled as either occupied or free based on laser measurements.

1.6 SIMULATION SOFTWARE

Advances in simulation platforms and virtual control environments [250], along with the availability of affordable computers with high processing power and fast graphics cards, and the growth of open-source programming communities have significantly accelerated the development of agricultural mobile robots. Simulation software and virtual environments are used as an affordable and reliable framework for experimenting with different sensing and acting mechanisms in order to verify the performance functionality of the robot in dynamic scenarios. The main drawbacks of robotic simulation in agriculture is that the real world always presents more complicated situations, such as unexpected disturbances to the actuators or unpredicted noise to the sensors feedback as a result of natural field conditions [250]. Figure 1.10 shows screenshots of a simulation study for the feasibility of using two different electrical tractors for autonomous mowing application in orchards [60]. There is a long list of academic and professional simulation platforms that can be adapted and used for agricultural robots. Some of the most widely used software are presented in Table 1.13, including Webots [251], Gazebo [252], and CoppeliaSim (previously known as V-REP) [251]. These software enable validating various aspects of mobile robot behavior and performance in a virtual environment before deploying them in agricultural fields.

Gazebo is an open-source robot simulation software widely used in the robotics community that offers a versatile platform with robust physics engine, high-quality graphics, and a variety of supported interfaces and sensors for simulating mobile robots and in a realistic and high-fidelity environment. Gazebo supports physics-based simulations including Open Dynamics Engine (ODE), Bullet, and Dynamic Animation and Robotics Toolkit (DART) to simulate rigid body dynamics. It utilizes the Object-Oriented Graphics Rendering Engine (OGRE) for high-performance rendering, enabling it to realistically model and simulate optical phenomena such as shadows, reflections, and refractions. Gazebo uses an XML format called Simulation Description Format (SDF) [259] that allows defining objects, environments, and robots in a structured manner for detailed configuration of the simulation elements and physics properties. Various sensors such as cameras, laser range finders, and

FIGURE 1.10 Screenshot of two simulated electrical tractors used for experimenting with collision avoidance sensors.

TABLE 1.13

Comparison Between General Specifications of Commonly Used Simulation Software for Agricultural Mobile Robots

Software Name	Developer	Supported OS	Physics Engine	Programming Language(s)	API Support	CAD File support	ROS Support	Release Year
Gazebo [252]	Open Robotics	Linux, macOS, Windows	ODE	C++, Python	C++, Python	STL, COLLADA (DAE), OBJ	Yes	2002
Webots [251]	Cyberbotics Ltd.	Windows, macOS, Linux	Proprietary physics engine	C, C++, Python, ROS	C, C++, ROS Python, Java, MATLAB®,	VRML, COLLADA (DAE)	Yes	1998
CoppeliaSim (V-REP) [251]	Coppelia Robotics	Windows, macOS, Linux	Proprietary physics engine	Lua, Python, MATLAB	Lua, Python, MATLAB, C, C++, ROS	OBJ, STL, COLLADA (DAE)	Limited	2010
MATLAB Robotics Toolbox [253]	MathWorks	Windows, macOS, Linux	MATLAB physics simulation	MATLAB	MATLAB	STEP, STL, VRML	Limited	N/A
Unity3D [254]	Unity Technologies	Windows, macOS, Linux	Unity Physics	C#, JavaScript	C#, JavaScript	FBX, OBJ, DAE	No	2005
Stage [255]	Willow Garage	Linux	Proprietary physics engine	C++, Python	C++, Python	N/A	Yes	N/A
PyRobotSim [256]	Open Source	Cross-platform	Pygame physics engine	Python	Python	N/A	Limited	N/A
Actin [250]	Energid Technologies	Windows, Linux	Proprietary physics engine	C++, Python	C++, Python	N/A	No	N/A
RoboDK [257]	RoboDK	Windows, macOS, Linux	Proprietary physics engine	Python, C#, ROS	Python, C#, ROS	STEP, IGES, STL, DXF, COLLADA (DAE), VRML	Yes	2014
ARGoS [258]	Institute for Systems and Robotics (ISR)	Windows, macOS, Linux	ODE	C++, Lua	C++, Lua	N/A	Limited	2004

IMUs, alongside the incorporation of noise to emulate real-world sensor inaccuracies and failures, are available in Gazebo. Gazebo also provides ROS and ROS 2 [260] integration, making it a popular choice for researchers and developers working with ROS-based robots and for effective communication between simulated robotic entities and algorithmic implementations.

CoppeliaSim [251], formerly known as V-REP (Virtual Robot Experimentation Platform), is a versatile and highly extensible simulator that operates on a distributed control architecture, allowing it to simultaneously handle various applications and devices in a synchronized manner. This architecture enhances its capacity to manage multiple robots, sensors, and actuators effectively within a single simulation environment. CoppeliaSim supports multiple physics engines such as Bullet, ODE, and Vortex to simulate various physical phenomena, such as rigid body dynamics, collisions, and friction, providing users with the flexibility to select an engine that best suits their application. Control of simulated robots can be achieved using both embedded scripting in the software's own scripting language, Lua, and external APIs such as Python, C++, ROS, and MATLAB®. CoppeliaSim also provides functionalities for creating detailed models and defining shape, size, kinematics, dynamics, and visual properties of robots and other entities in the simulation. Figure 1.11 shows screenshots of a simulation study for experimenting with different types of sensors for mapping and collision avoidance inside Gazebo and CoppeliaSim.

MATLAB's Robotics System Toolbox provides an interface and complete workflow to design, simulate, and deploy robotics algorithms and applications with ROS support that allow users to read and write data from and to ROS topics, service calls, and action servers, enabling real-time interaction with robots. Furthermore, it provides an array of robotic algorithms for autonomous navigation, including localization using particle filters, path planning via Rapidly-exploring Random Trees (RRT) [261], and sensor fusion and tracking functionalities, enabling the synthesis and testing of custom algorithms for mobile robots. The toolbox facilitates kinematic computations utilizing Denavit-Hartenberg (D-H) parameters [262], offering capabilities to model robotic manipulators and analyze joint trajectories, which is of great interest for validating control strategies in robotic arm applications.

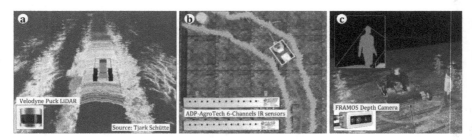

FIGURE 1.11   Screenshots of a simulation study on the use of (a) LiDAR sensor in Gazebo, (b) multi-arrays of distance detection sensors for collision avoidance in CoppeliaSim, and (c) depth camera for human detection in MATLAB® [106].

Webots [251] is another professional robot simulator developed by Cyberbotics that emphasizes ease of use, scalability, and a comprehensive library of robot models, making it a popular choice for both educational and industrial purposes. It offers a user-friendly interface and a library of pre-defined robot models, including various types of wheeled robots, legged robots, drones, and robotic arms, besides a flexible programming interface that supports multiple programming languages such as C, C++, Python, and Java. Webots is cross-platform compatible, supporting Windows, macOS, and Linux operating systems. It provides users with tools to model various sensors (i.e., cameras, LiDARs, GPS, IMUs) and actuators (i.e., motors and servos) for analyzing robot behaviors, control strategies, and algorithms. A notable feature of Webots is the remote control functionality and cloud-based simulations [263], enabling users to test their algorithms on actual hardware and utilize cloud computing resources for large-scale simulation studies, respectively.

Unity3D stands out as a powerful cross-platform software, offering a rich set of features to develop, simulate, and render 3D and 2D environments. Employing a component-based object system and C# scripting API, Unity enables developers to construct detailed physical and virtual environments, utilizing its physically based rendering (PBR) [264] capabilities to produce highly realistic visuals, along with a potent physics engine to accurately simulate real-world dynamics and interactions. Unity provides a suite of tools for immersive real-time simulations and visualizations, and it supports a vast array of file formats for 3D models and other assets, facilitating ease of integration with various design and modeling tools. Furthermore, Unity's extensible plugin architecture and robust APIs enable interfacing with different hardware devices, such as Virtual Reality (VR) and Augmented Reality (AR) headsets [265], sensors, and actuators, and it can be also interfaced with multiple third-party tools and platforms, like ROS, through custom middleware, expanding its applicability in autonomous mobile robots simulation. With its vast ecosystem and active developer community, Unity3D serves researchers and engineers in creating visually rich, interactive simulations and applications across diverse platforms and devices.

Stage [255] is a 2D robotics simulator that is particularly efficient for swarm robotics [266] and multi-robot system [267] testing, and is often interfaced with ROS to facilitate enhanced robot control and algorithm validation. PyRobotSim [256] is a Python-developed 3D robot simulation environment that adopts ease of algorithm iteration via Python scripts for promoting accessible robotic experimentation, particularly for beginners and educational endeavors. Actin [258] specializes in simulating robot kinematics and dynamics, providing a suitable platform for modeling and controlling complex robotic mechanisms, particularly manipulator arms, and excelling in real-time control, collision detection, and path planning in complex, obstacle-rich environments [268]. RoboDK [257] establishes itself as a pivotal tool for offline programming and simulation of industrial robots with an extensive library of robot models, which can subsequently be deployed to physical robot hardware without necessitating manual teaching. Lastly, ARGoS [258], with a focus on simulating large-scale robot swarms [269], provides a multi-physics simulation environment, allowing for the simultaneous simulation of heterogeneous robots [270] and optimizing computational resources by permitting the environment to be partitioned among different physics engines. Together, these platforms, each with its own particular emphasis and capabilities, support various sectors of agricultural mobile robots, research, and education, offering safe environments for the testing, analysis, and validation of different scenarios, sensors, and algorithms.

1.7 TECHNOLOGICAL CHALLENGES TOWARD COMMERCIALIZATION

Some of the main technical challenges toward commercialization of agricultural mobile robots include sensor accuracy, environmental conditions, issues with GPS denial environments (i.e., interference from physical objects, inconsistent and jamming GPS signals), data management and analytics, and integration with existing systems [271]. Robots are expected to accurately perceive their environment in real time for navigation inside large and complex farm environments, even in adverse weather conditions. This task, however, presents challenges from a sensing and perception standpoint since these robots are generally tested under specific environmental conditions, such as clear weather or level terrain. In challenging weather or rough terrain, the existing technology may struggle to execute tasks effectively, particularly in crops with intricate growth patterns or fields abundant in obstacles. Moreover, agricultural fields often feature natural and human-made obstructions such as trees, buildings, and other objects that can disrupt GPS signals. This interference can result in the robots losing GPS connectivity, hindering their navigation and task execution. In certain environments, GPS signals may prove inconsistent, leading to weak or fluctuating signals and subsequent navigation and task accuracy challenges. Furthermore, GPS signal inconsistencies can result in imprecise field data collection and mapping. Deliberate GPS signal jamming can occur in specific environments, posing significant security threats to agricultural mobile robots and other GPS-dependent systems. Jamming may be orchestrated by malicious actors or occur unintentionally due to other interference sources. Sensors generate copious amounts of data, demanding efficient management and analysis to inform farming decisions, which require advanced data management, analytics capabilities, and robust machine learning algorithms. These robots must seamlessly integrate with existing farming systems to maximize benefits for farmers. Achieving this necessitates robust and flexible software, alongside close collaboration between farmers and technology providers.

1.8 CONCLUSION

This chapter provided an overview of the sensing technology, algorithms, and software packages that are commonly used in the autonomy of agricultural mobile robots. In order to achieve widespread commercialization of these robots, there must be a clear and consistent regulatory framework in place, as well as standardized interfaces and protocols to ensure their safety and reliability for operating in fields and near crops. Depending on the technology used for the autonomous navigation and collision avoidance of agricultural mobile robots, regulations can be complex and restrictive, making it difficult for farmers to use these robots effectively. For some sensors, data collected about crops and soil conditions can be subjected to strict data privacy regulations that must be followed. In some regions, the commercial use of autonomous robots for farming might be limited, restricted, or even banned due to concerns about safety and environmental impact.

Aside from the technological solutions, investment and funding are playing a critical role in the commercialization of agricultural mobile robots, which can be expensive to purchase and maintain (with some models costing hundreds of thousands of dollars). This high cost can be a barrier for small and medium-sized farmers who may not have the financial resources to invest in and justify this technology. Private firms and venture capital companies are investing heavily in this field, with a focus on startups that are developing cutting-edge technologies and innovative solutions. Governments and industry organizations are also providing funding and support for research and development in order to promote the wider adoption of mobile robots for farming. The market for agricultural mobile robots is growing rapidly around the world, with different regions experiencing different levels of demands for these machines. According to a report by ResearchAndMarkets, the global agricultural robot market is expected to reach $7.98 billion by 2026, growing at a compound annual growth rate (CAGR) of 15.8% between 2021 and 2026. This growth is driven by a number of factors, including increasing demand for food, advances in technology, and the need for more sustainable and efficient farming methods. This market is highly competitive, with some of the key players including John Deere, Autonomous Tractor Corporation, AGCO Corporation, Case IH, CNH Industrial, Naio Technologies, Bosch Deepfield Robotics, and AgJunctio Kubota Corporation. These companies are investing heavily in research and development, as well as marketing and sales, in order to build their brand and increase their market share.

REFERENCES

[1]   L. Green, E. Webb, E. Johnson, S. Wynn, and C. Bogen, "Cost-Effective Approach to Explore Key Impacts on the Environment from Agricultural Tools to Inform Sustainability Improvements: Inversion Tillage as a Case Study," *Environmental Sciences Europe*, vol. 35, no. 1, p. 79, 2023.

[2]   A. Hossain et al., "Cost-Effective and Eco-Friendly Agricultural Technologies in Rice-Wheat Cropping Systems for Food and Environmental Security," in *Sustainable Intensification for Agroecosystem Services and Management*, M. K. Jhariya, A. Banerjee, R. S. Meena, S. Kumar, and A. Raj, Eds. Singapore: Springer, 2021, pp. 69–96.

[3]   L. Klerkx, E. Jakku, and P. Labarthe, "A Review of Social Science on Digital Agriculture, Smart Farming and Agriculture 4.0: New Contributions and a Future Research Agenda," *NJAS—Wageningen Journal of Life Sciences*, vol. 90–91, p. 100315, 2019.

[4]   H. N. Azmi, S. S. H. Hajjaj, K. R. Gsangaya, M. T. H. Sultan, M. F. Mail, and L. S. Hua, "Design and Fabrication of an Agricultural Robot for Crop Seeding," *Materials Today: Proceedings*, vol. 81, pp. 283–289, 2023.

[5]   S. Wang et al., "Design and Development of Orchard Autonomous Navigation Spray System," *Frontiers in Plant Science*, vol. 13, 2022.

[6]   V. Rajendran et al., "Towards Autonomous Selective Harvesting: A Review of Robot Perception, Robot Design, Motion Planning and Control," *Journal of Field Robotics*, 2023. https://doi.org/10.1002/rob.22230

[7] N. Virlet, K. Sabermanesh, P. Sadeghi-Tehran, and M. J. Hawkesford, "Field Scanalyzer: An Automated Robotic Field Phenotyping Platform for Detailed Crop Monitoring," *Functional Plant Biology*, vol. 44, no. 1, pp. 143–153, 2017.

[8] P. Baur and A. Iles, "Replacing Humans with Machines: A Historical Look at Technology Politics in California Agriculture," *Agriculture and Human Values*, vol. 40, no. 1, pp. 113–140, 2023.

[9] A. Bechar and C. Vigneault, "Agricultural Robots for Field Operations: Concepts and Components," *Biosystems Engineering*, vol. 149, pp. 94–111, 2016.

[10] F. B. P. Malavazi, R. Guyonneau, J.-B. Fasquel, S. Lagrange, and F. Mercier, "LiDAR-Only Based Navigation Algorithm for an Autonomous Agricultural Robot," *Computers and Electronics in Agriculture*, vol. 154, pp. 71–79, 2018.

[11] D. Ball et al., "Vision-Based Obstacle Detection and Navigation for an Agricultural Robot," *Journal of Field Robotics*, vol. 33, no. 8, pp. 1107–1130, 2016.

[12] Y. Li, M. Iida, T. Suyama, M. Suguri, and R. Masuda, "Implementation of Deep-Learning Algorithm for Obstacle Detection and Collision Avoidance for Robotic Harvester," *Computers and Electronics in Agriculture*, vol. 174, p. 105499, 2020.

[13] M. Bergerman, J. Billingsley, J. Reid, and E. van Henten, "Robotics in Agriculture and Forestry," in *Springer Handbook of Robotics*, B. Siciliano and O. Khatib, Eds. Cham: Springer International Publishing, 2016, pp. 1463–1492.

[14] S. G. Vougioukas, "Agricultural Robotics," *Annual Review of Control, Robotics, and Autonomous Systems*, vol. 2, pp. 365–392, 2019.

[15] W. L. Stone, *The History of Robotics*. Boca Raton, FL: CRC Press, 2018.

[16] S. Raikwar, J. Fehrmann, and T. Herlitzius, "Navigation and Control Development for a Four-Wheel-Steered Mobile Orchard Robot Using Model-Based Design," *Computers and Electronics in Agriculture*, vol. 202, p. 107410, 2022.

[17] L. Grimstad and P. J. From, "The Thorvald II Agricultural Robotic System," *Robotics*, vol. 6, no. 4. 2017.

[18] D. Sarri, S. Lombardo, R. Lisci, V. De Pascale, and M. Vieri, "AgroBot Smash a Robotic Platform for the Sustainable Precision Agriculture," in *Innovative Biosystems Engineering for Sustainable Agriculture, Forestry and Food Production*, Coppola, A., Di Renzo, G., Altieri, G., D'Antonio, P. Eds. MID-TERM AIIA 2019. Lecture Notes in Civil Engineering, vol 67. Springer, Cham. https://doi.org/10.1007/978-3-030-39299-4_85

[19] D. Ball et al., "Farm Workers of the Future: Vision-Based Robotics for Broad-Acre Agriculture," *IEEE Robotics & Automation Magazine*, vol. 24, no. 3, pp. 97–107, September 2017. https://doi.org/10.1109/MRA.2016.2616541

[20] G. A. Bekey, "On Autonomous Robots," *The Knowledge Engineering Review*, vol. 13, no. 2, pp. 143–146, 1998.

[21] L. Chapman, C. Gray, and C. Headleand, "A Sense-Think-Act Architecture for Low-Cost Mobile Robotics," in *Research and Development in Intelligent Systems XXXII*, Bramer, M., Petridis, M. Eds. SGAI 2015. Springer, Cham. https://doi.org/10.1007/978-3-319-25032-8_34

[22] F. Rubio, F. Valero, and C. Llopis-Albert, "A Review of Mobile Robots: Concepts, Methods, Theoretical Framework, and Applications," *International Journal of Advanced Robotic Systems*, vol. 16, no. 2, p. 1729881419839596, 2019.

[23] D. Bochtis, H. W. Griepentrog, S. Vougioukas, P. Busato, R. Berruto, and K. Zhou, "Route Planning for Orchard Operations," *Computers and Electronics in Agriculture*, vol. 113, pp. 51–60, 2015.

[24] T. Kröger, "Literature Survey: Trajectory Generation in and Control of Robotic Systems," in *On-Line Trajectory Generation in Robotic Systems: Basic Concepts*

for *Instantaneous Reactions to Unforeseen (Sensor) Events*, T. Kröger, Ed. Berlin, Heidelberg: Springer Berlin Heidelberg, 2010, pp. 11–31.

[25] M. N. A. Wahab, S. Nefti-Meziani, and A. Atyabi, "A Comparative Review on Mobile Robot Path Planning: Classical or Meta-Heuristic Methods?" *Annual Reviews in Control*, vol. 50, pp. 233–252, 2020.

[26] S. Zaman, L. Comba, A. Biglia, D. Ricauda Aimonino, P. Barge, and P. Gay, "Cost-Effective Visual Odometry System for Vehicle Motion Control in Agricultural Environments," *Computers and Electronics in Agriculture*, vol. 162, pp. 82–94, 2019.

[27] Y. Tang et al., "Recognition and Localization Methods for Vision-Based Fruit Picking Robots: A Review," *Frontiers in Plant Science*, vol. 11, 2020.

[28] F. Visentin et al., "A Mixed-Autonomous Robotic Platform for Intra-Row and Inter-Row Weed Removal for Precision Agriculture," *Computers and Electronics in Agriculture*, vol. 214, p. 108270, 2023.

[29] M. Perez-Ruiz, J. Martínez-Guanter, and S. K. Upadhyaya, "High-Precision GNSS for Agricultural Operations," in *GPS and GNSS Technology in Geosciences*, G. Srivastava, Ed. Amsterdam, The Netherlands: Elsevier, 2021, pp. 299–335. https://doi.org/10.1016/B978-0-12-818617-6.00017-2

[30] F. Rovira-Más, I. Chatterjee, and V. Sáiz-Rubio, "The Role of GNSS in the Navigation Strategies of Cost-Effective Agricultural Robots," *Computers and Electronics in Agriculture*, vol. 112, pp. 172–183, 2015.

[31] J. Cremona, R. Comelli, and T. Pire, "Experimental Evaluation of Visual-Inertial Odometry Systems for Arable Farming," *Journal of Field Robotics*, vol. 39, no. 7, pp. 1121–1135, 2022.

[32] F. Guo, H. Yang, X. Wu, H. Dong, Q. Wu, and Z. Li, "Model-Based Deep Learning for Low-Cost IMU Dead Reckoning of Wheeled Mobile Robot," *IEEE Transactions on Industrial Electronics*, pp. 1–11, 2023.

[33] T. Hague, J. A. Marchant, and N. D. Tillett, "Ground Based Sensing Systems for Autonomous Agricultural Vehicles," *Computers and Electronics in Agriculture*, vol. 25, no. 1, pp. 11–28, 2000.

[34] S. Wang et al., "Fusing Vegetation Index and Ridge Segmentation for Robust Vision Based Autonomous Navigation of Agricultural Robots in Vegetable Farms," *Computers and Electronics in Agriculture*, vol. 213, p. 108235, 2023.

[35] Y. Bai, B. Zhang, N. Xu, J. Zhou, J. Shi, and Z. Diao, "Vision-Based Navigation and Guidance for Agricultural Autonomous Vehicles and Robots: A Review," *Computers and Electronics in Agriculture*, vol. 205, p. 107584, 2023.

[36] F. Magistri et al., "Contrastive 3D Shape Completion and Reconstruction for Agricultural Robots Using RGB-D Frames," *IEEE Robotics and Automation Letters*, vol. 7, no. 4, pp. 10120–10127, 2022.

[37] T. Wang, B. Chen, Z. Zhang, H. Li, and M. Zhang, "Applications of Machine Vision in Agricultural Robot Navigation: A Review," *Computers and Electronics in Agriculture*, vol. 198, p. 107085, 2022.

[38] M. F. S. Xaud, A. C. Leite, and P. J. From, "Thermal Image Based Navigation System for Skid-Steering Mobile Robots in Sugarcane Crops," in *2019 International Conference on Robotics and Automation (ICRA)*, Montreal, QC, Canada, 2019, pp. 1808–1814. https://doi.org/10.1109/ICRA.2019.8794354

[39] A. Milella, G. Reina, and M. Nielsen, "A Multi-Sensor Robotic Platform for Ground Mapping and Estimation Beyond the Visible Spectrum," *Precision Agriculture*, vol. 20, no. 2, pp. 423–444, 2019.

[40] N. Tsoulias, M. Zhao, D. S. Paraforos, and D. Argyropoulos, "Hyper- and Multi-Spectral Imaging Technologies," in *Encyclopedia of Smart Agriculture Technologies*, Q. Zhang, Ed. Cham: Springer International Publishing, 2022, pp. 1–11.

[41]   G. Reina, A. Milella, R. Rouveure, M. Nielsen, R. Worst, and M. R. Blas, "Ambient Awareness for Agricultural Robotic Vehicles," *Biosystems Engineering*, vol. 146, pp. 114–132, 2016.

[42]   H. Ding, B. Zhang, J. Zhou, Y. Yan, G. Tian, and B. Gu, "Recent Developments and Applications of Simultaneous Localization and Mapping in Agriculture," *Journal of Field Robotics*, vol. 39, no. 6, pp. 956–983, 2022.

[43]   U. Weiss and P. Biber, "Plant Detection and Mapping for Agricultural Robots Using a 3D LIDAR Sensor," *Robotics and Autonomous Systems*, vol. 59, no. 5, pp. 265–273, 2011.

[44]   R. R. Shamshiri, C. Weltzien, and T. Schutte, "Multi-Sensor Data Fusion with Fuzzy Knowledge-Based Controller for Collision Avoidance of a Mobile Robot," in *LAND. TECHNIK 2022: The Forum for Agricultural Engineering Innovations*, 1st ed., VDI Wissensforum GmbH, Ed. Düsseldorf: VDI Verlag, 2022, pp. 349–358.

[45]   Q. Chen, H. Lin, J. Kuang, Y. Luo, and X. Niu, "Rapid Initial Heading Alignment for MEMS Land Vehicular GNSS/INS Navigation System," *IEEE Sensors Journal*, vol. 23, no. 7, pp. 7656–7666, 2023.

[46]   N. Van Nguyen, W. Cho, and K. Hayashi, "Performance Evaluation of a Typical Low-Cost Multi-Frequency Multi-GNSS Device for Positioning and Navigation in Agriculture—Part 1: Static Testing," *Smart Agricultural Technology*, vol. 1, p. 100004, 2021.

[47]   H. Wang, Z. Ma, Y. Ren, S. Du, H. Lu, Y. Shang, S. Hu, G. Zhang, Z. Meng, C. Wen, and W. Fu, "Interactive Image Segmentation Based Field Boundary Perception Method and Software for Autonomous Agricultural Machinery Path Planning," *Computers and Electronics in Agriculture*, vol. 217, p. 108568, 2024. https://doi.org/10.1016/j.compag.2023.108568

[48]   G. Fastellini and C. Schillaci, "Chapter 7—Precision Farming and IoT Case Studies Across the World," in *Agricultural Internet of Things and Decision Support for Precision Smart Farming*, A. Castrignanò, G. Buttafuoco, R. Khosla, A. M. Mouazen, D. Moshou, and O. Naud, Eds. Academic Press, 2020, pp. 331–415, ISBN 9780128183731. https://doi.org/10.1016/B978-0-12-818373-1.00007-X. (https://www.sciencedirect.com/science/article/pii/B978012818373100007X)

[49]   P. Visconti, F. Iaia, R. De Fazio, and N. I. Giannoccaro, "A Stake-Out Prototype System Based on GNSS-RTK Technology for Implementing Accurate Vehicle Reliability and Performance Tests," *Energies*, vol. 14, no. 16, 2021.

[50]   M. Bakken, V. R. Ponnambalam, R. J. D. Moore, J. G. O. Gjevestad, and P. J. From, "Robot-Supervised Learning of Crop Row Segmentation," in *2021 IEEE International Conference on Robotics and Automation (ICRA)*, Xi'an, China, 2021, pp. 2185–2191. https://doi.org/10.1109/ICRA48506.2021.9560815

[51]   J. Fernández-Novales, J. Tardáguila, S. Gutiérrez, and M. P. Diago, "On-the-go VIS + SW – NIR Spectroscopy as a Reliable Monitoring Tool for Grape Composition Within the Vineyard," *Molecules*, vol. 24, no. 15, 2019.

[52]   G. M. Sharipov et al., "Smart Implements by Leveraging ISOBUS: Development and Evaluation of Field Applications," *Smart Agricultural Technology*, vol. 6, p. 100341, 2023.

[53]   I. Kovács and I. Husti, "The Role of Digitalization in the Agricultural 4.0–How to Connect the Industry 4.0 to Agriculture?" *Hungarian Agricultural Engineering*, no. 33, pp. 38–42, 2018.

[54]   C. P. Baillie, C. R. Lobsey, D. L. Antille, C. L. McCarthy, and J. A. Thomasson, "A Review of the State of the Art in Agricultural Automation. Part III: Agricultural Machinery Navigation Systems," in *2018 ASABE Annual International Meeting*.

American Society of Agricultural and Biological Engineers, 2018, p. 1. https://elibrary. asabe.org/abstract.asp?aid=49511

[55]   C.-W. Jeon, H.-J. Kim, C. Yun, M.-S. Kang, and X. Z. Han, "Development of an Optimized Complete Infield-Path Planning Algorithm for an Autonomous Tillage Tractor in Polygonal Paddy Field and Validation Using 3D Virtual Tractor Simulator," in *2019 ASABE Annual International Meeting*. American Society of Agricultural and Biological Engineers, 2019, p. 1. https://elibrary.asabe.org/abstract.asp?aid=50559

[56]   S. S. Virk et al., "Case Study: Distribution Uniformity of a Blended Fertilizer Applied Using a Variable-Rate Spinner-Disc Spreader," *Applied Engineering in Agriculture*, vol. 29, no. 5, pp. 627–636, 2013.

[57]   T. Oksanen, R. Linkolehto, and I. Seilonen, "Adapting an Industrial Automation Protocol to Remote Monitoring of Mobile Agricultural Machinery: A Combine Harvester with IoT," *IFAC-PapersOnLine*, vol. 49, no. 16, pp. 127–131, 2016.

[58]   F. Radicioni, A. Stoppini, R. Brigante, A. Brozzi, and G. Tosi, "GNSS Network RTK for Automatic Guidance in Agriculture: Testing and Performance Evaluation," in *International Conference on Computational Science and Its Applications*. Cham: Springer International Publishing, 2020, pp. 19–35. https://link.springer.com/ chapter/10.1007/978-3-030-58814-4_2

[59]   M. Pérez-Ruiz, J. Carballido, J. Agüera, and J. A. Gil, "Assessing GNSS Correction Signals for Assisted Guidance Systems in Agricultural Vehicles," *Precision Agriculture*, vol. 12, no. 5, pp. 639–652, 2011.

[60]   C. Weltzien and R. R. Shamshiri, "SunBot: Autonomous Nursing Assistant for Emission-Free Berry Production, General Concepts and Framework," *Land.Technik AgEng*, pp. 463–470, 2019.

[61]   A. S. Aguiar, F. N. dos Santos, J. B. Cunha, H. Sobreira, and A. J. Sousa, "Localization and Mapping for Robots in Agriculture and Forestry: A Survey," *Robotics*, vol. 9, no. 4, 2020.

[62]   M. B. Alatise and G. P. Hancke, "A Review on Challenges of Autonomous Mobile Robot and Sensor Fusion Methods," *IEEE Access*, vol. 8, pp. 39830–39846, 2020.

[63]   K. Wang, S. Ma, J. Chen, F. Ren, and J. Lu, "Approaches, Challenges, and Applications for Deep Visual Odometry: Toward Complicated and Emerging Areas," *IEEE Transactions on Cognitive and Developmental Systems*, vol. 14, no. 1, pp. 35–49, 2022.

[64]   Y. Lu and D. Song, "Visual Navigation Using Heterogeneous Landmarks and Unsupervised Geometric Constraints," *IEEE Transactions on Robotics*, vol. 31, no. 3, pp. 736–749, 2015.

[65]   V. Moysiadis, N. Tsolakis, D. Katikaridis, C. G. Sørensen, S. Pearson, and D. Bochtis, "Mobile Robotics in Agricultural Operations: A Narrative Review on Planning Aspects," *Applied Sciences*, vol. 10, no. 10, p. 3453, 2020.

[66]   B. Kehoe, S. Patil, P. Abbeel, and K. Goldberg, "A Survey of Research on Cloud Robotics and Automation," *IEEE Transactions on Automation Science and Engineering*, vol. 12, no. 2, pp. 398–409, 2015.

[67]   H. Nehme, C. Aubry, T. Solatges, X. Savatier, R. Rossi, and R. Boutteau, "LiDAR-Based Structure Tracking for Agricultural Robots: Application to Autonomous Navigation in Vineyards," *Journal of Intelligent & Robotic Systems*, vol. 103, no. 4, p. 61, 2021.

[68]   G. Gil, D. E. Casagrande, L. P. Cortés, and R. Verschae, "Why the Low Adoption of Robotics in the Farms? Challenges for the Establishment of Commercial Agricultural Robots," *Smart Agricultural Technology*, vol. 3, p. 100069, 2023.

[69]   M. Almasri, K. Elleithy, and A. Alajlan, "Sensor Fusion Based Model for Collision Free Mobile Robot Navigation," *Sensors*, vol. 16, no. 1, 2016.

[70]   H.-M. Zhang, M.-L. Li, and L. Yang, "Safe Path Planning of Mobile Robot Based on Improved a* Algorithm in Complex Terrains," *Algorithms*, vol. 11, no. 4, 2018.

[71]   M. Skoczeń et al., "Obstacle Detection System for Agricultural Mobile Robot Application Using RGB-D Cameras," *Sensors*, vol. 21, no. 16, 2021.

[72]   K. Al-Muteb et al., "An Autonomous Stereovision-Based Navigation System (ASNS) for Mobile Robots," *Intelligent Service Robotics*, vol. 9, no. 3, pp. 187–205, 2016.

[73]   A. Wijanarko, A. P. Nugroho, L. Sutiarso, and T. Okayasu, "Development of Mobile Robovision with Stereo Camera for Automatic Crop Growth Monitoring in Plant Factory," *AIP Conference Proceedings*, vol. 2202, no. 1, p. 20100, 2019.

[74]   I. Papagianopoulos, G. De Mey, A. Kos, B. Wiecek, and V. Chatziathasiou, "Obstacle Detection in Infrared Navigation for Blind People and Mobile Robots," *Sensors*, vol. 23, no. 16, 2023.

[75]   F. Vulpi, R. Marani, A. Petitti, G. Reina, and A. Milella, "An RGB-D Multi-View Perspective for Autonomous Agricultural Robots," *Computers and Electronics in Agriculture*, vol. 202, p. 107419, 2022.

[76]   W.-S. Kim, D.-H. Lee, Y.-J. Kim, T. Kim, W.-S. Lee, and C.-H. Choi, "Stereo-Vision-Based Crop Height Estimation for Agricultural Robots," *Computers and Electronics in Agriculture*, vol. 181, p. 105937, 2021.

[77]   İ. Ünal, Ö. Kabaş, O. Eceoğlu, and G. Moiceanu, "Adaptive Multi-Robot Communication System and Collision Avoidance Algorithm for Precision Agriculture," *Applied Sciences*, vol. 13, no. 15, 2023.

[78]   N. Zou, Z. Xiang, and J. Zhang, "Multi-Spectrum Superpixel Based Obstacle Detection Under Vegetation Environments," in *2017 IEEE Intelligent Vehicles Symposium (IV)*, Los Angeles, CA, 2017, pp. 1209–1214. https://doi.org/10.1109/IVS.2017.7995877

[79]   K. Jakubczyk, B. Siemiątkowska, R. Więckowski, and J. Rapcewicz, "Hyperspectral Imaging for Mobile Robot Navigation," *Sensors*, vol. 23, no. 1, 2023.

[80]   K. Schueler, T. Weiherer, E. Bouzouraa, and U. Hofmann, "360 Degree Multi Sensor Fusion for Static and Dynamic Obstacles," in *2012 IEEE Intelligent Vehicles Symposium*, Madrid, Spain, 2012, pp. 692–697. https://doi.org/10.1109/IVS.2012.6232253

[81]   G. Adamides et al., "Design and Development of a Semi-Autonomous Agricultural Vineyard Sprayer: Human–Robot Interaction Aspects," *Journal of Field Robotics*, vol. 34, no. 8, pp. 1407–1426, 2017.

[82]   Y.-W. Choi, J.-W. Choi, S.-G. Im, D. Qian, and S.-G. Lee, "Multi-Robot Avoidance Control Based on Omni-Directional Visual SLAM with a Fisheye Lens Camera," *International Journal of Precision Engineering and Manufacturing*, vol. 19, no. 10, pp. 1467–1476, 2018.

[83]   D. Tiozzo Fasiolo, L. Scalera, E. Maset, and A. Gasparetto, "Recent Trends in Mobile Robotics for 3D Mapping in Agriculture," in *Advances in Service and Industrial Robotics. RAAD 2022. Mechanisms and Machine Science*, A. Müller and M. Brandstötter, Eds. Cham: Springer, 2022, vol. 120, pp. 428–435. https://doi.org/10.1007/978-3-031-04870-8_50

[84]   A. Ravankar, A. A. Ravankar, A. Rawankar, and Y. Hoshino, "Autonomous and Safe Navigation of Mobile Robots in Vineyard with Smooth Collision Avoidance," *Agriculture*, vol. 11, no. 10, 2021.

[85]   J. Taher, T. Hakala, A. Jaakkola, H. Hyyti, A. Kukko, P. Manninen, J. Maanpää, and J. Hyyppä, "Feasibility of Hyperspectral Single Photon Lidar for Robust Autonomous Vehicle Perception," *Sensors*, vol. 22, no. 15, p. 5759, 2022. https://doi.org/10.3390/s22155759

[86] C. Premachandra, S. Ueda, and Y. Suzuki, "Detection and Tracking of Moving Objects at Road Intersections Using a 360-Degree Camera for Driver Assistance and Automated Driving," *IEEE Access*, vol. 8, pp. 135652–135660, 2020.

[87] D. Tiozzo Fasiolo, L. Scalera, E. Maset, and A. Gasparetto, "Towards Autonomous Mapping in Agriculture: A Review of Supportive Technologies for Ground Robotics," *Robotics and Autonomous Systems*, vol. 169, p. 104514, 2023.

[88] N. Arago et al., "Smart Dairy Cattle Farming and In-Heat Detection Through the Internet of Things (IoT)," *International Journal of Integrated Engineering*, vol. 14, no. 1, pp. 157–172, 2022.

[89] R. Manish, Z. An, A. Habib, M. R. Tuinstra, and D. J. Cappelleri, "AgBug: Agricultural Robotic Platform for In-Row and Under Canopy Crop Monitoring and Assessment," in *Proceedings of the ASME 2021 International Design Engineering Technical Conferences and Computers and Information in Engineering Conference. Volume 8B: 45th Mechanisms and Robotics Conference (MR)*. Virtual, Online, 17 August 2021. V08BT08A017. ASME. https://doi.org/10.1115/DETC2021-68143

[90] Z. Kang and W. Zou, "Improving Accuracy of VI-SLAM with Fish-Eye Camera Based on Biases of Map Points," *Advanced Robotics*, vol. 34, no. 19, pp. 1272–1278, 2020.

[91] P. Schmidt, J. Scaife, M. Harville, S. Liman, and A. Ahmed, "Intel® RealSense™ Tracking Camera T265 and Intel® RealSense™ Depth Camera D435-Tracking and Depth," *Real Sense*, 2019. https://www.intelrealsense.com/wp-content/uploads/2019/11/Intel_RealSense_Tracking_and_Depth_Whitepaper_rev001.pdf

[92] Y. Han, A. Ali Mokhtarzadeh, and S. Xiao, "Novel Cartographer Using an OAK-D Smart Camera for Indoor Robots Location and Navigation," *Journal of Physics: Conference Series*, vol. 2467, no. 1, p. 12029, 2023.

[93] T.-M. Wang and Z.-C. Shih, "Measurement and Analysis of Depth Resolution Using Active Stereo Cameras," *IEEE Sensors Journal*, vol. 21, no. 7, pp. 9218–9230, 2021.

[94] E. Curto and H. Araujo, "An Experimental Assessment of Depth Estimation in Transparent and Translucent Scenes for Intel Realsense D415, SR305 and L515," *Sensors*, vol. 22, no. 19, 2022.

[95] M. Servi et al., "Metrological Characterization and Comparison of D415, D455, L515 Realsense Devices in the Close Range," *Sensors*, vol. 21, no. 22, 2021.

[96] V. Tadić et al., "Application of the ZED Depth Sensor for Painting Robot Vision System Development," *IEEE Access*, vol. 9, pp. 117845–117859, 2021.

[97] I. C. F. S. Condotta, T. M. Brown-Brandl, S. K. Pitla, J. P. Stinn, and K. O. Silva-Miranda, "Evaluation of Low-Cost Depth Cameras for Agricultural Applications," *Computers and Electronics in Agriculture*, vol. 173, p. 105394, 2020.

[98] M. Antico et al., "Postural Control Assessment Via Microsoft Azure Kinect DK: An Evaluation Study," *Computer Methods and Programs in Biomedicine*, vol. 209, p. 106324, 2021.

[99] A. Canepa, E. Ragusa, R. Zunino, and P. Gastaldo, "T-RexNet—A Hardware-Aware Neural Network for Real-Time Detection of Small Moving Objects," *Sensors*, vol. 21, no. 4, 2021.

[100] S. Gunturu, A. Munir, H. Ullah, S. Welch, and D. Flippo, "A Spatial AI-Based Agricultural Robotic Platform for Wheat Detection and Collision Avoidance," *AI*, vol. 3, no. 3, pp. 719–738, 2022.

[101] L. He, Y. Wang, S. Velipasalar, and M. C. Gursoy, "Human Detection Using Mobile Embedded Smart Cameras," in *2011 Fifth ACM/IEEE International Conference on Distributed Smart Cameras*, Ghent, Belgium, 2011, pp. 1–6. https://doi.org/10.1109/ICDSC.2011.6042924

[102] Y. Yamasaki, M. Morie, and N. Noguchi, "Development of a High-Accuracy Autonomous Sensing System for a Field Scouting Robot," *Computers and Electronics in Agriculture*, vol. 193, p. 106630, 2022.

[103] C. White and J. N. Gilmore, "Imagining the Thoughtful Home: Google Nest and Logics of Domestic Recording," *Critical Studies in Media Communication*, vol. 40, no. 1, pp. 6–19, 2023.

[104] Y. He, Q. He, S. Fang, and Y. Liu, "Precise Wireless Camera Localization Leveraging Traffic-Aided Spatial Analysis," *IEEE Transactions on Mobile Computing*, no. 01, pp. 1–13, 2023.

[105] J. Zhang, J. Wei, P. An, and X. Liu, "Novel Geometric Calibration Method for Pan-Tilt Camera with Single Control Point," *IEEE Access*, vol. 11, pp. 34175–34185, 2023.

[106] R. Shamshiri, C. Weltzien, I. Zytoon, and B. Sakal, "Evaluation of Laser and Infrared Sensors with CANBUS Communication for Collision Avoidance of a Mobile Robot," in *Proceedings International Conference on Agricultural Engineering. AgEng LAND. TECHNIK 2022*. Düsseldorf: VDI Verlag GmbH (0083-5560/978-3-18092406-9), 2022, pp. 121–130. https://www.vdi-nachrichten.com/shop/ageng-land-technik-2022/

[107] T. Stanescu, L. A. Sandru, and V. Dolga, "Studies Regarding Detection of Obstacles with Different Geometric Shape Using Parallax Ping Sensor," in *2015 IEEE 10th Jubilee International Symposium on Applied Computational Intelligence and Informatics*, Timisoara, Romania, 2015, pp. 455–458. https://doi.org/10.1109/SACI.2015.7208247. https://ieeexplore.ieee.org/abstract/document/7208247

[108] F. N. dos Santos, H. Sobreira, D. Campos, R. Morais, A. Paulo Moreira, and O. Contente, "Towards a Reliable Robot for Steep Slope Vineyards Monitoring," *Journal of Intelligent & Robotic Systems*, vol. 83, no. 3, pp. 429–444, 2016.

[109] Y. Wang, X. Li, J. Zhang, S. Li, Z. Xu, and X. Zhou, "Review of Wheeled Mobile Robot Collision Avoidance Under Unknown Environment," *Science Progress*, vol. 104, no. 3, p. 00368504211037771, 2021.

[110] O. Serrano-Pérez, M. G. Villarreal-Cervantes, J. C. González-Robles, and A. Rodríguez-Molina, "Meta-Heuristic Algorithms for the Control Tuning of Omnidirectional Mobile Robots," *Engineering Optimization*, vol. 52, no. 2, pp. 325–342, 2020.

[111] X. Dai, S. Long, Z. Zhang, and D. Gong, "Mobile Robot Path Planning Based on Ant Colony Algorithm with a* Heuristic Method," *Frontiers in Neurorobotics*, vol. 13, 2019.

[112] B. K. Patle, G. Babu L, A. Pandey, D. R. K. Parhi, and A. Jagadeesh, "A Review: On Path Planning Strategies for Navigation of Mobile Robot," *Defence Technology*, vol. 15, no. 4, pp. 582–606, 2019.

[113] K. Karur, N. Sharma, C. Dharmatti, and J. E. Siegel, "A Survey of Path Planning Algorithms for Mobile Robots," *Vehicles*, vol. 3, no. 3, pp. 448–468, 2021.

[114] D. Agarwal and P. S. Bharti, "Implementing Modified Swarm Intelligence Algorithm Based on Slime Moulds for Path Planning and Obstacle Avoidance Problem in Mobile Robots," *Applied Soft Computing*, vol. 107, p. 107372, 2021.

[115] X. Zhang, A. Liniger, and F. Borrelli, "Optimization-Based Collision Avoidance," *IEEE Transactions on Control Systems Technology*, vol. 29, no. 3, pp. 972–983, 2021.

[116] S. M. H. Rostami, A. K. Sangaiah, J. Wang, and X. Liu, "Obstacle Avoidance of Mobile Robots Using Modified Artificial Potential Field Algorithm," *EURASIP Journal on Wireless Communications and Networking*, vol. 2019, no. 1, pp. 1–19, 2019.

[117] P. Wang, S. Gao, L. Li, B. Sun, and S. Cheng, "Obstacle Avoidance Path Planning Design for Autonomous Driving Vehicles Based on an Improved Artificial Potential Field Algorithm," *Energies*, vol. 12, no. 12, 2019.

[118]   J. J. Kandathil, R. Mathew, and S. S. Hiremath, "Development and Analysis of a Novel Obstacle Avoidance Strategy for a Multi-Robot System Inspired by the Bug-1 Algorithm," *SIMULATION*, vol. 96, no. 10, 2020.

[119]   S. S, D. A. Nandesh, R. Raman K, G. K, and R. R, "An Investigation of Bug Algorithms for Mobile Robot Navigation and Obstacle Avoidance in Two-Dimensional Unknown Static Environments," *2021 International Conference on Communication Information and Computing Technology (ICCICT)*, Mumbai, India, 2021, pp. 1–6. https://doi.org/10.1109/ICCICT50803.2021.9510118

[120]   S. K. Das, K. Roy, T. Pandey, A. Kumar, A. K. Dutta, and S. K. Debnath, "Modified Critical Point—A Bug Algorithm for Path Planning and Obstacle Avoiding of Mobile Robot," in *2020 International Conference on Communication and Signal Processing (ICCSP)*, Chennai, India, 2020, pp. 351–356. https://doi.org/10.1109/ICCSP48568.2020.9182347

[121]   J. Li, M. Ran, H. Wang, and L. Xie, "A Behavior-Based Mobile Robot Navigation Method with Deep Reinforcement Learning," *Unmanned Systems*, vol. 9, no. 3, pp. 201–209, 2020.

[122]   B. Saeedi and M. Sadedel, "Implementation of Behavior-Based Navigation Algorithm on Four-Wheel Steering Mobile Robot," *Journal of Computational Applied Mechanics*, vol. 52, no. 4, pp. 619–641, 2021.

[123]   K. K. Pandey and D. R. Parhi, "Trajectory Planning and the Target Search by the Mobile Robot in an Environment Using a Behavior-Based Neural Network Approach," *Robotica*, vol. 38, no. 9, pp. 1627–1641, 2020.

[124]   M. Al-darwbi and U. Baroudi, "FreeD∗: A Mechanism for Finding a Short and Collision Free Path," *IET Cyber-Systems and Robotics*, vol. 1, no. 2, pp. 55–62, 2019.

[125]   Q. Wu, H. Lin, Y. Jin, Z. Chen, S. Li, and D. Chen, "A New Fallback Beetle Antennae Search Algorithm for Path Planning of Mobile Robots with Collision-Free Capability," *Soft Computing*, vol. 24, no. 3, pp. 2369–2380, 2020.

[126]   X. Zhong, J. Tian, H. Hu, and X. Peng, "Hybrid Path Planning Based on Safe A* Algorithm and Adaptive Window Approach for Mobile Robot in Large-Scale Dynamic Environment," *Journal of Intelligent & Robotic Systems*, vol. 99, no. 1, pp. 65–77, 2020.

[127]   M. Luo, X. Hou, and J. Yang, "Surface Optimal Path Planning Using an Extended Dijkstra Algorithm," *IEEE Access*, vol. 8, pp. 147827–147838, 2020.

[128]   S. Alshammrei, S. Boubaker, and L. Kolsi, "Improved Dijkstra Algorithm for Mobile Robot Path Planning and Obstacle Avoidance," *Computers, Materials and Continua*, vol. 72, pp. 5939–5954, 2022.

[129]   Z. Zhang, B. Qiao, W. Zhao, and X. Chen, "A Predictive Path Planning Algorithm for Mobile Robot in Dynamic Environments Based on Rapidly Exploring Random Tree," *Arabian Journal for Science and Engineering*, vol. 46, no. 9, pp. 8223–8232, 2021.

[130]   L. Wang, X. Yang, Z. Chen, and B. Wang, "Application of the Improved Rapidly Exploring Random Tree Algorithm to an Insect-Like Mobile Robot in a Narrow Environment," *Biomimetics*, vol. 8, no. 4, 2023.

[131]   J. Chen, Y. Zhao, and X. Xu, "Improved RRT-Connect Based Path Planning Algorithm for Mobile Robots," *IEEE Access*, vol. 9, pp. 145988–145999, 2021.

[132]   D. Wang, W. Wei, Y. Yeboah, Y. Li, and Y. Gao, "A Robust Model Predictive Control Strategy for Trajectory Tracking of Omni-Directional Mobile Robots," *Journal of Intelligent & Robotic Systems*, vol. 98, no. 2, pp. 439–453, 2020.

[133]   P. M. Blok, K. van Boheemen, F. K. van Evert, J. IJsselmuiden, and G.-H. Kim, "Robot Navigation in Orchards with Localization Based on Particle Filter and Kalman Filter," *Computers and Electronics in Agriculture*, vol. 157, pp. 261–269, 2019.

[134] M. Lv, H. Wei, X. Fu, W. Wang, and D. Zhou, "A Loosely Coupled Extended Kalman Filter Algorithm for Agricultural Scene-Based Multi-Sensor Fusion," *Frontiers in Plant Science*, vol. 13, 2022.

[135] S. Erfani, A. Jafari, and A. Hajiahmad, "Comparison of Two Data Fusion Methods for Localization of Wheeled Mobile Robot in Farm Conditions," *Artificial Intelligence in Agriculture*, vol. 1, pp. 48–55, 2019.

[136] T. Sprodowski, "Collision Avoidance for Mobile Robots Based on an Occupancy Grid," in *Recent Advances in Model Predictive Control: Theory, Algorithms, and Applications*, T. Faulwasser, M. A. Müller, and K. Worthmann, Eds. Cham: Springer International Publishing, 2021, pp. 219–244.

[137] C. Patruno, V. Renò, N. Mosca, M. di Summa, and M. Nitti, "A Robust Method for 2D Occupancy Map Building for Indoor Robot Navigation," *Proc. SPIE*, vol. 11785, p. 117850D, 2021.

[138] J. Zhang, X. Wang, L. Xu, and X. Zhang, "An Occupancy Information Grid Model for Path Planning of Intelligent Robots," *ISPRS International Journal of Geo-Information*, vol. 11, no. 4. 2022.

[139] F. H. Ajeil, I. K. Ibraheem, A. T. Azar, and A. J. Humaidi, "Grid-Based Mobile Robot Path Planning Using Aging-Based Ant Colony Optimization Algorithm in Static and Dynamic Environments," *Sensors*, vol. 20, no. 7. 2020.

[140] H. K. Tripathy, S. Mishra, H. K. Thakkar, and D. Rai, "CARE: A Collision-Aware Mobile Robot Navigation in Grid Environment Using Improved Breadth First Search," *Computers & Electrical Engineering*, vol. 94, p. 107327, 2021.

[141] S. Jiang, S. Wang, Z. Yi, M. Zhang, and X. Lv, "Autonomous Navigation System of Greenhouse Mobile Robot Based on 3D Lidar and 2D Lidar SLAM," *Frontiers in Plant Science*, vol. 13, 2022.

[142] W. Zhang, L. Gong, S. Huang, S. Wu, and C. Liu, "Factor Graph-Based High-Precision Visual Positioning for Agricultural Robots with Fiducial Markers," *Computers and Electronics in Agriculture*, vol. 201, p. 107295, 2022.

[143] Y. Yang et al., "An Optimal Goal Point Determination Algorithm for Automatic Navigation of Agricultural Machinery: Improving the Tracking Accuracy of the Pure Pursuit Algorithm," *Computers and Electronics in Agriculture*, vol. 194, p. 106760, 2022.

[144] L. Wang, Z. Zhai, Z. Zhu, and E. Mao, "Path Tracking Control of an Autonomous Tractor Using Improved Stanley Controller Optimized with Multiple-Population Genetic Algorithm," *Actuators*, vol. 11, no. 1, 2022.

[145] Y. Sun, B. Cui, F. Ji, X. Wei, and Y. Zhu, "The Full-Field Path Tracking of Agricultural Machinery Based on PSO-Enhanced Fuzzy Stanley Model," *Applied Sciences*, vol. 12, no. 15, 2022.

[146] H. Chen, X. Wang, L. Zhao, R. Jiang, and B. Zhumadil, "Research on Path Tracking Control of Mobile Storage Robot Based on Model Predictive Control and Linear Quadratic Regulator," *Proc. SPIE*, vol. 12748, p. 127482O, 2023.

[147] W. Fu, Y. Liu, and X. Zhang, "Research on Accurate Motion Trajectory Control Method of Four-Wheel Steering AGV Based on Stanley-PID Control," *Sensors*, vol. 23, no. 16. 2023.

[148] X. Jin, S.-L. Dai, J. Liang, D. Guo, and H. Tan, "Constrained Line-of-Sight Tracking Control of a Tractor-Trailer Mobile Robot System with Multiple Constraints," in *2021 American Control Conference (ACC)*, New Orleans, LA, 2021, pp. 1046–1051. https://doi.org/10.23919/ACC50511.2021.9483365

[149] J. Shu, H. Liu, J. Dong, P. Du, and W. Zhang, "Trajectory Tracking Control of Differential Driving Double-Wheeled Mobile Robot," in *2022 IEEE 2nd International Conference*

on *Electronic Technology, Communication and Information (ICETCI)*, Changchun, China, 2022, pp. 943–947. https://doi.org/10.1109/ICETCI55101.2022.9832119

[150] F. Liu, X. Li, S. Yuan, and W. Lan, "Slip-Aware Motion Estimation for Off-Road Mobile Robots Via Multi-Innovation Unscented Kalman Filter," *IEEE Access*, vol. 8, pp. 43482–43496, 2020.

[151] S. O. H. Madgwick, S. Wilson, R. Turk, J. Burridge, C. Kapatos, and R. Vaidyanathan, "An Extended Complementary Filter for Full-Body MARG Orientation Estimation," *IEEE/ASME Transactions on Mechatronics*, vol. 25, no. 4, pp. 2054–2064, 2020.

[152] M. Liu, Y. Cai, L. Zhang, and Y. Wang, "Attitude Estimation Algorithm of Portable Mobile Robot Based on Complementary Filter," *Micromachines*, vol. 12, no. 11, 2021.

[153] C. Yang, D. Wang, Y. Zeng, Y. Yue, and P. Siritanawan, "Knowledge-Based Multimodal Information Fusion for Role Recognition and Situation Assessment by Using Mobile Robot," *Information Fusion*, vol. 50, pp. 126–138, 2019.

[154] M. F. Aslan, A. Durdu, A. Yusefi, K. Sabanci, and C. Sungur, "A Tutorial: Mobile Robotics, SLAM, Bayesian Filter, Keyframe Bundle Adjustment and ROS Applications," in *Robot Operating System (ROS): The Complete Reference (Volume 6)*, A. Koubaa, Ed. Cham: Springer International Publishing, 2021, pp. 227–269.

[155] B. J. Gogoi and P. K. Mohanty, "A Braitenberg Path Planning Strategy for E-puck Mobile Robot in Simulation and Real-Time Environments," in *Applications of Computational Methods in Manufacturing and Product Design*. Lecture Notes in Mechanical Engineering, B. B. V. L. Deepak, D. Parhi, B. Biswal, and P. C. Jena, Eds. Singapore: Springer, 2022, pp. 135–151. https://doi.org/10.1007/978-981-19-0296-3_13

[156] D. H. Lee, S. S. Lee, C. K. Ahn, P. Shi, and C.-C. Lim, "Finite Distribution Estimation-Based Dynamic Window Approach to Reliable Obstacle Avoidance of Mobile Robot," *IEEE Transactions on Industrial Electronics*, vol. 68, no. 10, pp. 9998–10006, 2020.

[157] T. Hossain, H. Habibullah, R. Islam, and R. V. Padilla, "Local Path Planning for Autonomous Mobile Robots by Integrating Modified Dynamic-Window Approach and Improved Follow the Gap Method," *Journal of Field Robotics*, vol. 39, no. 4, pp. 371–386, 2022.

[158] J. N. Greenberg and X. Tan, "Dynamic Optical Localization of a Mobile Robot Using Kalman Filtering-Based Position Prediction," *IEEE/ASME Transactions on Mechatronics*, vol. 25, no. 5, pp. 2483–2492, 2020.

[159] M. Kheirandish, E. A. Yazdi, H. Mohammadi, and M. Mohammadi, "A Fault-Tolerant Sensor Fusion in Mobile Robots Using Multiple Model Kalman Filters," *Robotics and Autonomous Systems*, vol. 161, p. 104343, 2023.

[160] N. Akai, "Reliable Monte Carlo Localization for Mobile Robots," *Journal of Field Robotics*, vol. 40, no. 3, pp. 595–613, 2023.

[161] L. Xu, C. Feng, V. R. Kamat, and C. C. Menassa, "An Occupancy Grid Mapping Enhanced Visual SLAM for Real-Time Locating Applications in Indoor GPS-Denied Environments," *Automation in Construction*, vol. 104, pp. 230–245, 2019.

[162] J. Pak, J. Kim, Y. Park, and H. I. Son, "Field Evaluation of Path-Planning Algorithms for Autonomous Mobile Robot in Smart Farms," *IEEE Access*, vol. 10, pp. 60253–60266, 2022.

[163] L. Zhang et al., "A Quadratic Traversal Algorithm of Shortest Weeding Path Planning for Agricultural Mobile Robots in Cornfield," *Journal of Robotics*, vol. 2021, p. 6633139, 2021.

[164] B. Li, C. Dong, Q. Chen, Y. Mu, Z. Fan, Q. Wang, and X. Chen, "Path Planning of Mobile Robots Based on an Improved A*algorithm," in *Proceedings of the 2020 4th High Performance Computing and Cluster Technologies Conference & 2020 3rd International Conference on Big Data and Artificial Intelligence (HPCCT & BDAI '20)*.

New York: Association for Computing Machinery, 2020, pp. 49–53. https://doi. org/10.1145/3409501.3409524

[165] J. Kim and H. I. Son, "A Voronoi Diagram-Based Workspace Partition for Weak Cooperation of Multi-Robot System in Orchard," *IEEE Access*, vol. 8, pp. 20676–20686, 2020.

[166] M. R. Wayahdi, S. H. N. Ginting, and D. Syahputra, "Greedy, A-Star, and Dijkstra's Algorithms in Finding Shortest Path," *International Journal of Advances in Data and Information Systems*, vol. 2, no. 1 SE-, pp. 45–52, 2021.

[167] X. Wang, X. Luo, B. Han, Y. Chen, G. Liang, and K. Zheng, "Collision-Free Path Planning Method for Robots Based on an Improved Rapidly-Exploring Random Tree Algorithm," *Applied Sciences*, vol. 10, no. 4, 2020.

[168] S. Chakraborty, D. Elangovan, P. L. Govindarajan, M. F. ELnaggar, M. M. Alrashed, and S. Kamel, "A Comprehensive Review of Path Planning for Agricultural Ground Robots," *Sustainability*, vol. 14, no. 15, 2022.

[169] H. Sun, W. Zhang, R. Yu, and Y. Zhang, "Motion Planning for Mobile Robots—Focusing on Deep Reinforcement Learning: A Systematic Review," *IEEE Access*, vol. 9, pp. 69061–69081, 2021.

[170] L. Chang, L. Shan, C. Jiang, and Y. Dai, "Reinforcement Based Mobile Robot Path Planning with Improved Dynamic Window Approach in Unknown Environment," *Autonomous Robots*, vol. 45, no. 1, pp. 51–76, 2021.

[171] X. Ji, S. Feng, Q. Han, H. Yin, and S. Yu, "Improvement and Fusion of A* Algorithm and Dynamic Window Approach Considering Complex Environmental Information," *Arabian Journal for Science and Engineering*, vol. 46, no. 8, pp. 7445–7459, 2021.

[172] V. Vasilopoulos, Y. Kantaros, G. J. Pappas, and D. E. Koditschek, "Reactive Planning for Mobile Manipulation Tasks in Unexplored Semantic Environments," in *2021 IEEE International Conference on Robotics and Automation (ICRA)*, Xi'an, China, 2021, pp. 6385–6392. https://doi.org/10.1109/ICRA48506.2021.9561958

[173] A. Tuncer and M. Yildirim, "Dynamic Path Planning of Mobile Robots with Improved Genetic Algorithm," *Computers & Electrical Engineering*, vol. 38, no. 6, pp. 1564–1572, 2012.

[174] N. H. Amer, H. Zamzuri, K. Hudha, and Z. A. Kadir, "Modelling and Control Strategies in Path Tracking Control for Autonomous Ground Vehicles: A Review of State of the Art and Challenges," *Journal of Intelligent & Robotic Systems*, vol. 86, no. 2, pp. 225–254, 2017.

[175] M. Samuel, M. Hussein, and M. B. Mohamad, "A Review of Some Pure-Pursuit Based Path Tracking Techniques for Control of Autonomous Vehicle," *International Journal of Computer Applications*, vol. 135, no. 1, pp. 35–38, 2016.

[176] R. K. Mandava and P. R. Vundavilli, "An Adaptive PID Control Algorithm for the Two-Legged Robot Walking on a Slope," *Neural Computing and Applications*, vol. 32, no. 8, pp. 3407–3421, 2020.

[177] C. Wang, X. Liu, X. Yang, F. Hu, A. Jiang, and C. Yang, "Trajectory Tracking of an Omni-Directional Wheeled Mobile Robot Using a Model Predictive Control Strategy," *Applied Sciences*, vol. 8, no. 2, 2018.

[178] Z. Liu, Y. Zhang, C. Yuan, L. Ciarletta, and D. Theilliol, "Collision Avoidance and Path Following Control of Unmanned Aerial Vehicle in Hazardous Environment," *Journal of Intelligent & Robotic Systems*, vol. 95, no. 1, pp. 193–210, 2019.

[179] N. Razmjooy, M. Ramezani, and A. Namadchian, "A New LQR Optimal Control for a Single-Link Flexible Joint Robot Manipulator Based on Grey Wolf Optimizer," *Majlesi Journal of Electrical Engineering*, vol. 10, no. 3, p. 53, 2016.

[180] M. Kamel, T. Stastny, K. Alexis, and R. Siegwart, "Model Predictive Control for Trajectory Tracking of Unmanned Aerial Vehicles Using Robot Operating System," in *Robot Operating System (ROS): The Complete Reference (Volume 2)*, A. Koubaa, Ed. Cham: Springer International Publishing, 2017, pp. 3–39.

[181] E. Kelasidi, P. Liljebäck, K. Y. Pettersen, and J. T. Gravdahl, "Integral Line-of-Sight Guidance for Path Following Control of Underwater Snake Robots: Theory and Experiments," *IEEE Transactions on Robotics*, vol. 33, no. 3, pp. 610–628, 2017.

[182] Á. Madridano, A. Al-Kaff, D. Martín, and A. de la Escalera, "Trajectory Planning for Multi-Robot Systems: Methods and Applications," *Expert Systems with Applications*, vol. 173, p. 114660, 2021.

[183] Z. Shiller, "Off-Line and On-Line Trajectory Planning," in *Motion and Operation Planning of Robotic Systems: Background and Practical Approaches*, G. Carbone and F. Gomez-Bravo, Eds. Cham: Springer International Publishing, 2015, pp. 29–62.

[184] J. Kim and E. A. Croft, "Online Near Time-Optimal Trajectory Planning for Industrial Robots," *Robotics and Computer-Integrated Manufacturing*, vol. 58, pp. 158–171, 2019.

[185] G. F. Welch, "Kalman Filter," in *Computer Vision: A Reference Guide*. Cham: Springer International Publishing, 2020, pp. 1–3.

[186] M. B. Alatise and G. P. Hancke, "Pose Estimation of a Mobile Robot Based on Fusion of IMU Data and Vision Data Using an Extended Kalman Filter," *Sensors*, vol. 17, no. 10. 2017.

[187] J. Fayyad, M. A. Jaradat, D. Gruyer, and H. Najjaran, "Deep Learning Sensor Fusion for Autonomous Vehicle Perception and Localization: A Review," *Sensors*, vol. 20, no. 15, 2020.

[188] J. Gui, D. Gu, S. Wang, and H. Hu, "A Review of Visual Inertial Odometry from Filtering and Optimisation Perspectives," *Advanced Robotics*, vol. 29, no. 20, pp. 1289–1301, 2015.

[189] X. Xu et al., "A Review of Multi-Sensor Fusion SLAM Systems Based on 3D LIDAR," *Remote Sensing*, vol. 14, no. 12, 2022.

[190] W. McAllister, J. Whitman, J. Varghese, A. Davis, and G. Chowdhary, "Agbots 3.0: Adaptive Weed Growth Prediction for Mechanical Weeding Agbots," *IEEE Transactions on Robotics*, vol. 38, no. 1, pp. 556–568, 2022.

[191] A. Bender, B. Whelan, and S. Sukkarieh, "A High-Resolution, Multimodal Data Set for Agricultural Robotics: A Ladybird's-Eye View of Brassica," *Journal of Field Robotics*, vol. 37, no. 1, pp. 73–96, 2020.

[192] S. Sotnik and V. Lyashenko, "Agricultural Robotic Platforms," *International Journal of Engineering and Information Systems (IJEAIS)*, vol. 6, no. 4, pp. 14–21, 2022. https://openarchive.nure.ua/entities/publication/b5c49781-bc60-491d-8127-48c1165e0316

[193] R. Xu and C. Li, "A Review of High-Throughput Field Phenotyping Systems: Focusing on Ground Robots," *Plant Phenomics*, vol. 2022, 2023.

[194] S. Panpatte and C. Ganeshkumar, "Artificial Intelligence in Agriculture Sector: Case Study of Blue River Technology," in *Proceedings of the Second International Conference on Information Management and Machine Intelligence*. Lecture Notes in Networks and Systems, vol. 166, D. Goyal, A. K. Gupta, V. Piuri, M. Ganzha, and M. Paprzycki, Eds. Singapore: Springer, 2021, pp. 147–153. https://doi.org/10.1007/978-981-15-9689-6_17

[195] D. Alejo, J. A. Cobano, G. Heredia, and A. Ollero, "A Reactive Method for Collision Avoidance in Industrial Environments," *Journal of Intelligent & Robotic Systems*, vol. 84, no. 1, pp. 745–758, 2016.

[196] F. Ingrand and M. Ghallab, "Deliberation for Autonomous Robots: A Survey," *Artificial Intelligence*, vol. 247, pp. 10–44, 2017.

[197] B. J. Gogoi and P. K. Mohanty, "Path Planning of E-puck Mobile Robots Using Braitenberg Algorithm," in *International Conference on Artificial Intelligence and Sustainable Engineering*. Lecture Notes in Electrical Engineering, vol. 837, G. Sanyal, C. M. Travieso-González, S. Awasthi, C. M. Pinto, and B. R. Purushothama, Eds. Singapore: Springer, 2022, pp. 139–150. https://doi.org/10.1007/978-981-16-8546-0_13

[198] D. Q. Bao and I. Zelinka, "Obstacle Avoidance for Swarm Robot Based on Self-Organizing Migrating Algorithm," *Procedia Computer Science*, vol. 150, pp. 425–432, 2019.

[199] J.-H. Liang and C.-H. Lee, "Efficient Collision-Free Path-Planning of Multiple Mobile Robots System Using Efficient Artificial Bee Colony Algorithm," *Advances in Engineering Software*, vol. 79, pp. 47–56, 2015.

[200] J. Iqbal, R. Xu, S. Sun, and C. Li, "Simulation of an Autonomous Mobile Robot for LiDAR-Based in-Field Phenotyping and Navigation," *Robotics*, vol. 9, no. 2, 2020.

[201] X. Xiao, B. Liu, G. Warnell, and P. Stone, "Motion Planning and Control for Mobile Robot Navigation Using Machine Learning: A Survey," *Autonomous Robots*, vol. 46, no. 5, pp. 569–597, 2022.

[202] Y. Li, M. Li, J. Qi, D. Zhou, Z. Zou, and K. Liu, "Detection of Typical Obstacles in Orchards Based on Deep Convolutional Neural Network," *Computers and Electronics in Agriculture*, vol. 181, p. 105932, 2021.

[203] N. Guo, C. Li, D. Wang, Y. Song, G. Liu, and T. Gao, "Local Path Planning of Mobile Robot Based on Long Short-Term Memory Neural Network," *Automatic Control and Computer Sciences*, vol. 55, no. 1, pp. 53–65, 2021.

[204] G. Farias, G. Garcia, G. Montenegro, E. Fabregas, S. Dormido-Canto, and S. Dormido, "Reinforcement Learning for Position Control Problem of a Mobile Robot," *IEEE Access*, vol. 8, pp. 152941–152951, 2020.

[205] A. Konar, I. G. Chakraborty, S. J. Singh, L. C. Jain, and A. K. Nagar, "A Deterministic Improved Q-Learning for Path Planning of a Mobile Robot," *IEEE Transactions on Systems, Man, and Cybernetics: Systems*, vol. 43, no. 5, pp. 1141–1153, 2013.

[206] X. Zhang, Y. Guo, J. Yang, D. Li, Y. Wang, and R. Zhao, "Many-Objective Evolutionary Algorithm Based Agricultural Mobile Robot Route Planning," *Computers and Electronics in Agriculture*, vol. 200, p. 107274, 2022.

[207] S. Zhang, Y. Pang, and G. Hu, "Trajectory-Tracking Control of Robotic System via Proximal Policy Optimization," in *2019 IEEE International Conference on Cybernetics and Intelligent Systems (CIS) and IEEE Conference on Robotics, Automation and Mechatronics (RAM)*, Bangkok, Thailand, 2019, pp. 380–385. https://doi.org/10.1109/CIS-RAM47153.2019.9095849

[208] S. Lin et al., "Automatic Detection of Plant Rows for a Transplanter in Paddy Field Using Faster R-CNN," *IEEE Access*, vol. 8, pp. 147231–147240, 2020.

[209] S. Chen and N. Noguchi, "Remote Safety System for a Robot Tractor Using a Monocular Camera and a YOLO-Based Method," *Computers and Electronics in Agriculture*, vol. 215, p. 108409, 2023.

[210] P. Estefo, J. Simmonds, R. Robbes, and J. Fabry, "The Robot Operating System: Package Reuse and Community Dynamics," *Journal of Systems and Software*, vol. 151, pp. 226–242, 2019.

[211] H. R. Kam, S.-H. Lee, T. Park, and C.-H. Kim, "RViz: A Toolkit for Real Domain Data Visualization," *Telecommunication Systems*, vol. 60, no. 2, pp. 337–345, 2015.

[212] L. Jaillet, J. Cortés, and T. Siméon, "Sampling-Based Path Planning on Configuration-Space Costmaps," *IEEE Transactions on Robotics*, vol. 26, no. 4, pp. 635–646, 2010.

[213] M. M. Almasri, A. M. Alajlan, and K. M. Elleithy, "Trajectory Planning and Collision Avoidance Algorithm for Mobile Robotics System," *IEEE Sensors Journal*, vol. 16, no. 12, pp. 5021–5028, 2016.

[214] A. Harchowdhury, L. Kleeman, and L. Vachhani, "High Density 3D Sensing Using a Nodding 2D LIDAR and Reconfigurable Mirrors," *Mechatronics*, vol. 92, p. 102968, 2023.

[215] Y. Y. Ziggah, H. Youjian, X. Yu, and L. P. Basommi, "Capability of Artificial Neural Network for Forward Conversion of Geodetic Coordinates (ϕ,λ,h) to Cartesian Coordinates (X, Y, Z)," *Mathematical Geosciences*, vol. 48, no. 6, pp. 687–721, 2016.

[216] K. P. Akhil, G. Manikutty, R. Ravindran, and B. Rao R, "Autonomous Navigation of an Unmanned Ground Vehicle for Soil Pollution Monitoring," in *2019 2nd International Conference on Intelligent Computing, Instrumentation and Control Technologies (ICICICT)*, Kannur, India, 2019, vol. 1, pp. 1563–1567. https://doi.org/10.1109/ICICICT46008.2019.8993292

[217] M. A. Post, A. Bianco, and X. T. Yan, "Autonomous Navigation with ROS for a Mobile Robot in Agricultural Fields," in *14th International Conference on Informatics in Control, Automation and Robotics (ICINCO), 2017-07-26–2017-07-28*. Universidad Rey Juan Carlos, 2017.

[218] J. Iqbal, R. Xu, H. Halloran, and C. Li, "Development of a Multi-Purpose Autonomous Differential Drive Mobile Robot for Plant Phenotyping and Soil Sensing," *Electronics*, vol. 9, no. 9, 2020.

[219] K. Zheng, "ROS Navigation Tuning Guide," in *Robot Operating System (ROS): The Complete Reference (Volume 6)*, A. Koubaa, Ed. Cham: Springer International Publishing, 2021, pp. 197–226.

[220] I. Ullah, Y. Shen, X. Su, C. Esposito, and C. Choi, "A Localization Based on Unscented Kalman Filter and Particle Filter Localization Algorithms," *IEEE Access*, vol. 8, pp. 2233–2246, 2020.

[221] W. A. S. Norzam, H. F. Hawari, and K. Kamarudin, "Analysis of Mobile Robot Indoor Mapping Using GMapping Based SLAM with Different Parameter," *IOP Conference Series: Materials Science and Engineering*, vol. 705, no. 1, p. 12037, 2019.

[222] X. Zhang, J. Lai, D. Xu, H. Li, and M. Fu, "2D Lidar-Based SLAM and Path Planning for Indoor Rescue Using Mobile Robots," *Journal of Advanced Transportation*, vol. 2020, p. 8867937, 2020.

[223] L. Zhi and M. Xuesong, "Navigation and Control System of Mobile Robot Based on ROS," in *2018 IEEE 3rd Advanced Information Technology, Electronic and Automation Control Conference (IAEAC)*, Chongqing, China, 2018, pp. 368–372. https://doi.org/10.1109/IAEAC.2018.8577901

[224] A. Filotheou, E. Tsardoulias, A. Dimitriou, A. Symeonidis, and L. Petrou, "Quantitative and Qualitative Evaluation of ROS-Enabled Local and Global Planners in 2D Static Environments," *Journal of Intelligent & Robotic Systems*, vol. 98, no. 3, pp. 567–601, 2020.

[225] P. Marin-Plaza, A. Hussein, D. Martin, and A. de la Escalera, "Global and Local Path Planning Study in a ROS-Based Research Platform for Autonomous Vehicles," *Journal of Advanced Transportation*, vol. 2018, p. 6392697, 2018.

[226] H. Do Quang et al., "An Approach to Design Navigation System for Omnidirectional Mobile Robot Based on ROS," *International Journal of Mechanical Engineering and Robotics Research*, vol. 9, no. 11, pp. 1502–1508, 2020.

[227] B. Gurevin et al., "A Novel GUI Design for Comparison of ROS-Based Mobile Robot Local Planners," *IEEE Access*, vol. 11, pp. 125738–125748, 2023.

[228] S. Pütz, J. S. Simón, and J. Hertzberg, "Move Base Flex a Highly Flexible Navigation Framework for Mobile Robots," in *2018 IEEE/RSJ International Conference on Intelligent Robots and Systems (IROS)*, Madrid, Spain, 2018, pp. 3416–3421. https://doi.org/10.1109/IROS.2018.8593829

[229] A. H. Abbas, M. I. Habelalmateen, S. Jurdi, L. Audah, and N. A. M. Alduais, "GPS Based Location Monitoring System with Geo-Fencing Capabilities," *AIP Conference Proceedings*, vol. 2173, no. 1, p. 20014, 2019.

[230] M. Z. Rodriguez et al., "Clustering Algorithms: A Comparative Approach," *PLoS ONE*, vol. 14, no. 1, p. e0210236, 2019.

[231] M. Hahsler, M. Piekenbrock, and D. Doran, "Dbscan: Fast Density-Based Clustering with R," *Journal of Statistical Software*, vol. 91, no. 1 SE-Articles, pp. 1–30, 2019.

[232] Y. Ji, S. Li, C. Peng, H. Xu, R. Cao, and M. Zhang, "Obstacle Detection and Recognition in Farmland Based on Fusion Point Cloud Data," *Computers and Electronics in Agriculture*, vol. 189, p. 106409, 2021.

[233] J. Martínez-Gómez, V. Morell, M. Cazorla, and I. García-Varea, "Semantic Localization in the PCL Library," *Robotics and Autonomous Systems*, vol. 75, pp. 641–648, 2016.

[234] D. Tardioli, R. Parasuraman, and P. Ögren, "Pound: A Multi-Master ROS Node for Reducing Delay and Jitter in Wireless Multi-Robot Networks," *Robotics and Autonomous Systems*, vol. 111, pp. 73–87, 2019.

[235] H. Si, J. Qiu, and Y. Li, "A Review of Point Cloud Registration Algorithms for Laser Scanners: Applications in Large-Scale Aircraft Measurement," *Applied Sciences*, vol. 12, no. 20, 2022.

[236] C. Kolhatkar and K. Wagle, "Review of SLAM Algorithms for Indoor Mobile Robot with LIDAR and RGB-D Camera Technology," in *Innovations in Electrical and Electronic Engineering*. Lecture Notes in Electrical Engineering, vol. 661, M. N. Favorskaya, S. Mekhilef, R. K. Pandey, and N. Singh, Eds. Singapore: Springer, 2021, pp. 397–409. https://doi.org/10.1007/978-981-15-4692-1_30

[237] M. A. S. Teixeira, H. B. Santos, A. S. de Oliveira, L. V. Arruda, and F. Neves, "Robots Perception Through 3D Point Cloud Sensors," in *Robot Operating System (ROS): The Complete Reference (Volume 2)*, A. Koubaa, Ed. Cham: Springer International Publishing, 2017, pp. 525–561.

[238] M. H. Saleem, J. Potgieter, and K. M. Arif, "Automation in Agriculture by Machine and Deep Learning Techniques: A Review of Recent Developments," *Precision Agriculture*, vol. 22, no. 6, pp. 2053–2091, 2021.

[239] B. Pang, E. Nijkamp, and Y. N. Wu, "Deep Learning with TensorFlow: A Review," *Journal of Educational and Behavioral Statistics*, vol. 45, no. 2, pp. 227–248, 2019.

[240] A. Paszke et al., "Pytorch: An Imperative Style, High-Performance Deep Learning Library," *Advances in Neural Information Processing Systems*, vol. 32, 2019.

[241] D. Holz, A. E. Ichim, F. Tombari, R. B. Rusu, and S. Behnke, "Registration with the Point Cloud Library: A Modular Framework for Aligning in 3-D," *IEEE Robotics & Automation Magazine*, vol. 22, no. 4, pp. 110–124, 2015.

[242] M. Labbé and F. Michaud, "RTAB-Map as an Open-Source Lidar and Visual Simultaneous Localization and Mapping Library for Large-Scale and Long-Term Online Operation," *Journal of Field Robotics*, vol. 36, no. 2, pp. 416–446, 2019.

[243] D.-X. Zhou, "Theory of Deep Convolutional Neural Networks: Downsampling," *Neural Networks*, vol. 124, pp. 319–327, 2020.

[244] M. Avzayesh, M. Abdel-Hafez, M. AlShabi, and S. A. Gadsden, "The Smooth Variable Structure Filter: A Comprehensive Review," *Digital Signal Processing*, vol. 110, p. 102912, 2021.

[245]   S. Qi et al., "Review of Multi-View 3D Object Recognition Methods Based on Deep Learning," *Displays*, vol. 69, p. 102053, 2021.

[246]   B. Zhao, J. Feng, X. Wu, and S. Yan, "A Survey on Deep Learning-Based Fine-Grained Object Classification and Semantic Segmentation," *International Journal of Automation and Computing*, vol. 14, no. 2, pp. 119–135, 2017.

[247]   O. I. Abiodun et al., "Comprehensive Review of Artificial Neural Network Applications to Pattern Recognition," *IEEE Access*, vol. 7, pp. 158820–158846, 2019.

[248]   S. Pütz, T. Wiemann, and J. Hertzberg, "Tools for Visualizing, Annotating and Storing Triangle Meshes in ROS and RViz," in *2019 European Conference on Mobile Robots (ECMR)*, Prague, Czech Republic, 2019, pp. 1–6. https://doi.org/10.1109/ECMR.2019.8870953

[249]   P. D. C. Cheng, M. Indri, F. Sibona, M. De Rose, and G. Prato, "Dynamic Path Planning of a Mobile Robot Adopting a Costmap Layer Approach in ROS2," in *2022 IEEE 27th International Conference on Emerging Technologies and Factory Automation (ETFA)*, Stuttgart, Germany, 2022, pp. 1–8. https://doi.org/10.1109/ETFA52439.2022.9921458

[250]   R. R. Shamshiri, I. A. Hameed, L. Pitonakova, C. Weltzien, S. K. Balasundram, I. J. Yule et al., "Simulation Software and Virtual Environments for Acceleration of Agricultural Robotics: Features Highlights and Performance Comparison," *International Journal of Agricultural and Biological Engineering*, vol. 11, no. 4, pp. 15–31, 2018.

[251]   A. Farley, J. Wang, and J. A. Marshall, "How to Pick a Mobile Robot Simulator: A Quantitative Comparison of CoppeliaSim, Gazebo, MORSE and Webots with a Focus on Accuracy of Motion," *Simulation Modelling Practice and Theory*, vol. 120, p. 102629, 2022.

[252]   F. Furrer, M. Burri, M. Achtelik, and R. Siegwart, "RotorS—A Modular Gazebo MAV Simulator Framework," in *Robot Operating System (ROS): The Complete Reference (Volume 1)*, A. Koubaa, Ed. Cham: Springer International Publishing, 2016, pp. 595–625.

[253]   S. Karakaya, G. Kucukyildiz, and H. Ocak, "A New Mobile Robot Toolbox for MATLAB," *Journal of Intelligent & Robotic Systems*, vol. 87, no. 1, pp. 125–140, 2017.

[254]   Y. Li, S. Dai, Y. Shi, L. Zhao, and M. Ding, "Navigation Simulation of a Mecanum Wheel Mobile Robot Based on an Improved A* Algorithm in Unity3D," *Sensors*, vol. 19, no. 13, 2019.

[255]   A. Ligot and M. Birattari, "Simulation-Only Experiments to Mimic the Effects of the Reality Gap in the Automatic Design of Robot Swarms," *Swarm Intelligence*, vol. 14, no. 1, pp. 1–24, 2020.

[256]   A. Sakai, D. Ingram, J. Dinius, K. Chawla, A. Raffin, and A. Paques, "Pythonrobotics: A Python Code Collection of Robotics Algorithms," *arXiv preprint arXiv:1808.10703*, 2018.

[257]   S. Chakraborty and P. S. Aithal, "Forward and Inverse Kinematics Demonstration Using RoboDK and C," *International Journal of Applied Engineering and Management Letters (IJAEML)*, vol. 5, no. 1, pp. 97–105, 2021.

[258]   L. Pitonakova, M. Giuliani, A. Pipe, and A. Winfield, "Feature and Performance Comparison of the V-REP, Gazebo and ARGoS Robot Simulators," in *Towards Autonomous Robotic Systems. TAROS 2018. Lecture Notes in Computer Science*, vol. 10965, M. Giuliani, T. Assaf, and M. Giannaccini, Eds. Cham: Springer, 2018, pp. 357–368. https://doi.org/10.1007/978-3-319-96728-8_30

[259] S. Kim, M. Peavy, P.-C. Huang, and K. Kim, "Development of BIM-Integrated Construction Robot Task Planning and Simulation System," *Automation in Construction*, vol. 127, p. 103720, 2021.

[260] S. Macenski, T. Foote, B. Gerkey, C. Lalancette, and W. Woodall, "Robot Operating System 2: Design, Architecture, and Uses in the Wild," *Science Robotics*, vol. 7, no. 66, p. eabm6074, 2022.

[261] I. Pérez-Hurtado, M. Á. Martínez-del-Amor, G. Zhang, F. Neri, and M. J. Pérez-Jiménez, "A Membrane Parallel Rapidly-Exploring Random Tree Algorithm for Robotic Motion Planning," *Integrated Computer-Aided Engineering*, vol. 27, no. 2, pp. 121–138, 2020.

[262] R. S. Othayoth, R. G. Chittawadigi, R. P. Joshi, and S. K. Saha, "Robot Kinematics Made Easy Using RoboAnalyzer Software," *Computer Applications in Engineering Education*, vol. 25, no. 5, pp. 669–680, 2017.

[263] S. J. E. Taylor et al., "The CloudSME Simulation Platform and Its Applications: A Generic Multi-Cloud Platform for Developing and Executing Commercial Cloud-Based Simulations," *Future Generation Computer Systems*, vol. 88, pp. 524–539, 2018.

[264] M. Pharr, W. Jakob, and G. Humphreys, *Physically Based Rendering: From Theory to Implementation*. Cambridge, MA: MIT Press, 2023.

[265] M. D. Moniruzzaman, A. Rassau, D. Chai, and S. M. S. Islam, "Teleoperation Methods and Enhancement Techniques for Mobile Robots: A Comprehensive Survey," *Robotics and Autonomous Systems*, vol. 150, p. 103973, 2022.

[266] A. R. Cheraghi, S. Shahzad, and K. Graffi, "Past, Present, and Future of Swarm Robotics," in *Intelligent Systems and Applications. IntelliSys 2021.* Lecture Notes in Networks and Systems, vol. 296, K. Arai, Ed. Cham: Springer, 2022, pp. 190–233. https://doi.org/10.1007/978-3-030-82199-9_13

[267] C. Ju, J. Kim, J. Seol, and H. Il Son, "A Review on Multirobot Systems in Agriculture," *Computers and Electronics in Agriculture*, vol. 202, p. 107336, 2022.

[268] P. K. Pedapati, S. K. Pradhan, and S. Kumar, "Nonlinear Adaptive Control of an Autonomous Ground Vehicle in Obstacle Rich Environment: Some Experimental Results," in *TENCON 2019–2019 IEEE Region 10 Conference (TENCON)*, Kochi, India, 2019, pp. 614–619. https://doi.org/10.1109/TENCON.2019.8929625

[269] M. Schranz, M. Umlauft, M. Sende, and W. Elmenreich, "Swarm Robotic Behaviors and Current Applications," *Frontiers in Robotics and AI*, p. 36, 2020.

[270] C. Ju and H. I. Son, "Modeling and Control of Heterogeneous Agricultural Field Robots Based on Ramadge–Wonham Theory," *IEEE Robotics and Automation Letters*, vol. 5, no. 1, pp. 48–55, 2020.

[271] A. Bechar and C. Vigneault, "Agricultural Robots for Field Operations. Part 2: Operations and Systems," *Biosystems Engineering*, vol. 153, pp. 110–128, 2017.

2 Robot-Assisted Soil Apparent Electrical Conductivity Measurements in Orchards

*Dimitrios Chatziparaschis, Elia Scudiero,
and Konstantinos Karydis*

2.1 INTRODUCTION

Agricultural geophysics employs non-invasive sensing techniques to characterize soil spatial variability and provide valuable insights into soil-plant-management relationships [1, 2]. Specifically, geospatial information on soil characteristics can indicate optimal cultivation approaches and may provide an estimate of expected yields [3, 4]. Soil salinity (i.e., salt content) is a crucial metric used to describe the soil characteristics and water content of an area. As such, it has been used widely across applications such as in agriculture, water management, geological mapping, and engineering surveys [5–7]. Information about bulk density, minerals content, pH, soil temperature, and more can be evaluated by measuring the apparent electrical conductivity (ECa) of the field and generating a profile of the surveyed land. Thus, approximating the ECa spatial variability of a field can provide a broader understanding of the water flow through the ground, pinpoint any spots with irregular soil patterns, and indicate the necessity of supplying additive plant nutrients or different irrigation approaches [8].

In general, soil conductivity is mainly estimated in-situ, using three distinctive methods [9]: moisture meters (hydrometers) installed into the ground, time-domain reflectometers, and measurement of soil electromagnetic induction (EMI). In the first two cases, growers install and use decentralized sensor arrays to gather information from selected points on the field [10–12]. A main drawback of these approaches is that they provide discrete measurements and hence sparse information over the complete field, which may lead to less efficient agricultural tactics and higher costs while aiming to scale over larger fields. On the other side, ECa is measured geospatially (on-the-go) with electrical resistivity methods and EMI sensors. The EMI measurements of soil apparent electrical conductivity can be performed in a continuous manner and proximally, whereby a farm worker walks through the field holding the EMI sensor or a field vehicle that carries the sensor is driven around [5] (Figure 2.1). In this way, a more spatially dense belief about field irrigation is formed as the sensor can gather either continuous measurements of selected regions within the field or sparse measurements from specific points. Figure 2.1a depicts an instance from a manual survey of soil moisture in an olive tree field, with the use of the GF CMD-Tiny EMI instrument. Despite the benefits afforded by continuous EMI soil measurements, a noteworthy drawback is that such surveys may not scale well in larger fields as they can become quite labor-intensive depending on the broader area of inspection and weather conditions (e.g., heat fatigue). The standard of practice (besides manual operation) is to use an ATV (all-terrain vehicle, Figure 2.1b) that carries a (often larger) sensor; however, due to its size, it may be hard to get the sensor close to the tree roots where estimating soil apparent electrical conductivity is most crucial. A robot-assisted solution has the potential to get the sensor close to the roots (as in the manual case) in a less laborious way (as in the ATV case) and thus bridge the gap between the two main standard of practice methods by offering an attractive alternative.

We gratefully acknowledge the support of USDA-NIFA grant # 2021-67022-33453, ONR grant # N00014-19-1-2252, the University of California under grant UC-MRPI M21PR3417, a Frank G. and Janice B. Delfino Agricultural Technology Research Initiative Seed Award, and an OASIS-IFA (Opportunities to Advance Sustainability, Innovation, and Social Inclusion—Internal Funding Awards). Any opinions, findings, and conclusions or recommendations expressed in this material are those of the authors and do not necessarily reflect the views of the funding agencies.

The use of (mobile) robots has been offering key assistance in contemporary survey and agronomy processes, for example, to better understand field conditions with the use of onboard sensors, provide real-time field modeling and decision-making on agricultural tactics, and in some cases offer aid to farm workers (e.g., transporting workers across the field or elevating them to reach parts of the tree that are high off the ground [13]). In many scenarios, unmanned ground vehicles (UGVs) are used to collect data from the ground [14–16], while other studies utilize aerial data taken by unmanned aerial vehicles (UAVs) [17–19] and make decisions [20, 21] for the surveying area on-the-fly. As a follow-up, other research demonstrates collaborative approaches with both aerial and ground robots [22–24], which can yield an even more detailed and broader inspection of the field. Out of all types of data collected in the field, the most relevant to this present work concerns soil moisture.

During the past years, there have been various approaches aiming to utilize robots for collection of soil moisture measurements. Some examples of related research include mobile robots conducting soil potential of hydrogen (pH) measurements for determination of soil health in the survey site [25], robot manipulators with onboard depth cameras that place soil moisture sensors on the plants [26], and even a UAV-based multispectral system for estimating and modeling soil salinity of the inspected field [27]. Thayer *et al.* [28] revealed the NP-hard routing problem of autonomous robots in precision irrigation scenarios and developed two domain-specific heuristic approaches.

FIGURE 2.1 (a) Instance of the manual data collection process using the handheld EMI sensor. Manually collected data serve as ground truth in this work. (b) The robot considered in this work is a Clearpath Jackal UGV wheeled robot, carrying a GF CMD-Tiny EMI instrument (long orange cylinder) for ECa measurements alongside a Polaris ATV. (c) GNSS positioning information for the robot to navigate autonomously as well as for geo-localization and cross-reference of obtained sensor measurements is provided via an RTK-Base station.

In more detail, Pulido Fentanes *et al.* [29] demonstrated a mobile robot that performs soil moisture measurements autonomously in the field, using a cosmic-ray sensor. The main drawback of using such sensory equipment is the increase in the overall cost of the application, which might be accurate but financially inefficient for some farming applications. On the aerial side, Tseng *et al.* [30] showed the usability of machine learning with aerial imagery to learn and predict the soil moisture conditions on individual plants and large fields through the air. Even though this application reported reduced water consumption by up to 52%, there is a need for a more accurate and robust system for measuring soil moisture levels. In addition, Lukowska *et al.* [31] presented an all-terrain six-wheeled robot, which utilizes a drilling system with a flap design to perform soil sampling in larger fields. Along similar lines, Dimaya *et al.* [32] developed a mobile soil robot collector (SoilBot) to automate soil collection in a sugar cane field. In such cases, as the post-processing of the soil samples will provide detailed information about the moisture level of the sampled soil, there may be approaches for real-time and broader field monitoring robotic applications. In addition, *Agrobot* [14] is a farm robot that has been designed to reduce human labor through autonomous seed sowing and soil moisture measuring. Further, Bourgeois *et al.* [33] demonstrated a low-cost and portable land robot, namely RoSS, for soil quality sensing purposes, which can collect soil samples and/ or insert soil sensor probes at sampling locations to measure the moisture levels. In these approaches, vital information for the total salinity of the field might be missing as local information is obtained from the water content in the soil of the sampled areas. Campbell *et al.* [34] showcased a small-factor robot with an attached EMI sensor for field-scale soil apparent electrical conductivity measurements. Even though this work presents an efficient approach toward integrating an EMI sensor onto a mobile robot to conduct continuous ECa measurements, it has limited traversability and operational time because of its small size.

In this study, we present a mid-sized ground mobile robot solution that is able to conduct semi-autonomous and on-demand continuous EMI ECa measurements under various and larger field environments, and obtain a field-scale ECa map. Our proposed solution is based on the Clearpath Jackal UGV, which is equipped with a customized and adjustable platform that can carry the EMI instrument GF CMD-Tiny. The robot supports teleoperation via Bluetooth, it can directly navigate through sending desired waypoints and can execute trajectories as it utilizes its onboard GPS and RTK positioning data along with local and global planners to reach desired goals. The design, hardware and software system integration, and testing of this platform are fully presented and evaluated in both simulated and real field-scale scenarios, including over bare fields and in muddy terrains. The proposed robot platform demonstrates high efficiency in terms of portability, traversability, as well as data collection since the robot is found to be capable of collecting data with high linearity compared to handheld (no-robot) cases. Thus, our proposed solution shows promise to serve as a useful tool in modern field survey, ECa mapping, and irrigation scheduling.

In the remainder of the chapter, we first discuss key components (off-the-shelf as well as fabricated in-house) and the overall system design, and offer key system integration information (Section 2.2). An important contribution of this work is a thorough study of the tradeoffs regarding EMI sensor placement on the mobile robot that involves tools and processes we develop for both experimental testing and testing in simulation; this is presented in Section 2.3. Details and key findings from an initial testing phase to validate the preliminary efficacy of the overall system are also reported in that section. Full system evaluation and testing results across multiple trials in two distinctive fields are presented in Section 2.4. Finally, Section 2.5 concludes this chapter.

2.2 SYSTEM DESIGN AND INTEGRATION OF KEY COMPONENTS

2.2.1 SOIL CONDUCTIVITY AND EMPLOYED SENSOR

Electromagnetic induction can help measure the soil apparent electrical conductivity of a field. The main operating principle is based on the evaluation of the induced magnetic field from the ground as transmitted by an electromagnetic conductivity meter. Specifically, the EMI transmitter emits a harmonic signal toward the ground and generates a magnetic field. Since the receiver is placed with the same dipole orientation as the transmitter, it captures the secondary (induced) magnetic field, which relates to the ground conductivity, namely out-of-phase measured in *mS/m* and the in-phase that is a relative metric to the primary magnetic field and measures the magnetic susceptibility of the area.

In our study, we use the CMD-Tiny meter from GF Instruments, which features a compact and lightweight build. This instrument has a control unit module that is used for configuring and logging the EMI measurements and the CMD probe that is the main magnetic sensing module. The latter component has a cylindrical shape with 50 *cm* length and 4.25 *cm* of diameter, and the total setup weighs 425 *g* (Figure 2.1a). This EMI instrument can obtain soil conductivity measurements from 0.35 *m* up to 0.7 *m* in-ground depth according to its setup and selected resolution.

2.2.2 MOBILE ROBOT SETUP

We use the Clearpath Robotics Jackal robot platform,[1] which is a UGV designed for use in outdoor and rugged all-terrain environments. The Jackal has been used in agricultural robotics research [35, 36], autonomous exploration [37–39], as well as social-aware navigation [40, 41]. The robot's dimensions are 50.8 × 43.2 × 25.4 *cm* with a payload area of 43 × 32.25 *cm* for mounting various (OEM and custom) onboard modules. Its available payload capacity reaches 20 *kg*. The robot features an onboard NVIDIA Jetson AGX Xavier computer that is responsible for all onboard computation. On the sensors and actuators side, the robot is equipped with a GNSS receiver, an IMU module, and motorized wheel encoders (besides the custom payloads developed in this work and which we discuss later, or additional sensors like stereo cameras and LiDAR that are routinely deployed on the robot for autonomous navigation [42, 43]). Additionally, the Jackal is an ROS-compatible robot, as it uses the ROS navigation stack and the ROS environment for its main functionality. The total operating time of this robot can reach up to 4 *hrs* depending on the use and type of the operating environment. Importantly, the required operating time on the field can vary depending on field size and type, and the desired field mapping resolution. For instance, the surveys conducted herein took place over a total period of 1.5 *hrs* (three surveys each lasting for 0.5 *hrs*), in a 30 × 15 *m* field, and with the robot moving with a stable linear speed of 1 *m/s*. Under these settings, the total battery consumption never exceeded 40% during our experiments in each case, for performing the total ECa mapping of the field. The operational time of the Jackal can be expanded by the use of a secondary/additional battery. If an increased amount of surveying time and operation in even larger fields are desired, an alternative commercial wheeled robot can be employed instead (e.g., the Clearpath Robotics Husky,[2] and Warthog UGV,[3] or the Amiga from Farm-ng)[4]; our method can directly apply to such robots as well. Figure 2.1b depicts the Jackal robot equipped with various onboard sensors, in comparison to a Polaris ATV that is typically used for mechanized EMI measurements.

2.2.3 POSITIONING SYSTEM INTEGRATION FOR FIELD NAVIGATION

Nowadays, high-accuracy global navigation satellite systems (GNSS) are used (along with field sensors) in precision agriculture to generate, extract, and obtain field observations with spatial information. In many cases, real-time kinematic (RTK) positioning is applied to provide *cm*-level accuracy on the captured data by using real-time corrections through an established and calibrated base station.

Herein we use the Holybro H-RTK F9P GNSS series as the high-accuracy positioning module for our field experiments, which integrates a differential high-precision multi-band GNSS positioning system. By selecting specific points on the field, we calibrate the RTK base station to obtain its position at *cm*-level, and we use telemetry to establish communication with the robot's onboard autopilot hardware. Figure 2.1c illustrates an H-RTK F9P RTK base station establishment in an outdoor environment. On the robot side, we use the Holybro Pixhawk 4 autopilot module in the rover airframe (UGV mode), along with the Holybro SiK Telemetry V3 100 *Mhz* and the GNSS receiver to capture the RTK corrections from the base station. The MAVROS software[5] is utilized to parse the positioning data and use them with the onboard computer through a USB connection at a rate of 10 *Hz*. In this way, the Jackal robot is able to reliably georeference every captured measurement in the field.

On the navigation side, the robot's captured positions are described in the World Geodetic System 1984 (WGS-84). Additionally, the Jackal uses an extended Kalman Filter [44], which fuses information from the onboard IMU and the wheel encoders to provide the state estimation of the robot (odometry). Since in our case we require the Jackal to be able to follow a predefined GNSS-tagged trajectory, we utilize the *navsat_transform_node* from the ROS navigation stack. Through this approach, initially, all of the requested geotagged targets are transformed to the Universal Transverse Mercator (UTM) coordinate system. Given this information and through the positioning data of the robot's odometry, a static transformation is generated to describe both the UTM coordinates of the robot and the targets' position into the local robot's frame. In this way, the requested trajectory can be georeferenced and transformed into Jackal's local frame, and thus the robot can follow it. It is worth mentioning that, in case the robot gets into a position where there is limited satellite visibility (i.e., GPS-denied environments), it continues

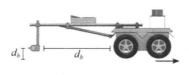

FIGURE 2.2 CAD rendering of the Jackal's platform to hold the GF CMD-Tiny instrument. The parameters d_h and d_b indicate the adjustable height and distance of the sensor probe, respectively.

geotagging the captured measurements based on pose belief estimation via fusion of the onboard IMU and wheel encoders odometry data.[6] While not employed in this work, our method is autonomous-ready in the sense that it can be directly integrated with waypoint navigation determined readily by onboard sensors (e.g., [45]), where task allocation and motion planning for (newly perceived) obstacle avoidance can happen online (e.g., [46, 47]). Also, mapped modalities from the flora [48] and the fauna [49] during the survey can even enhance and provide a multi-modal belief about the field conditions. Additionally, on the traversing side, we focus on open-world navigation without the existence of obstructive objects in the path that may cause the robot to get out of track to bypass them. The robot can get over small tree branches or crops as a farm worker would do, but the ground should be relatively uncluttered for the ECa inspection, as in a typical survey scenario. Also, as Jackal is an IP62 rugged robot, it has a capable high torque 4×4 drivetrain, allowing it to navigate through muddy and uneven parts similar to a common ATV. In our experiments, we test our proposed system in both dry and muddy environments to demonstrate our system's efficacy. The robot supports a 20 *cm* wheel diameter (up to 30 *cm*) and a variety of tire models, such as square (plastic) spike sets, which can make it even more versatile for challenging terrain types that require higher traction.

2.2.4 ROBOT CONFIGURATION AND THE DESIGN OF THE SENSOR MOUNTING PLATFORM

As our aim is to attach the CMD-Tiny meter on the Jackal robot, we start with the design of the sensor platform and the definition of its adjustable parameters. The prototype renderings are depicted in Figure 2.2. As the Jackal robot has a limited payload area, we aim to design and use a module that can be attached on the robot's top plate and extend outwards so that the sensor can get in close proximity to the ground. In this way, the robot can carry the EMI sensor and obtain soil conductivity measurements while navigating in a continuous manner. However, the sensor's sensitivity in magnetic field measurements to other metallic and/or electronic components (like the robot itself) in view of module placement on the robot requires a study of some key tradeoffs.

In particular, the EMI measurements can be distorted by the presence of other magnetic fields caused by the robot, as well as by random oscillations caused by the robot's movement. Thus, one intuitive solution would be to place the sensor as far as possible from the robot (i.e., large value for parameter d_b) and as close to the ground as possible (i.e., small value for parameter d_h). However, as distance d_b increases (Figure 2.2), the moment arm of the payload increases, which in turn can lead to higher power consumption of the robot. In addition, the magnitude of the oscillations (which directly affects EMI measurement consistency) will also increase. Finally, the longer the distance, the worse the capability of the robot to traverse uneven terrain and/or negotiate dips and bumps that are bound to exist in the field in practice, as possible sensor collisions with the ground can occur if height d_h is not sufficiently large. For these reasons, we design an adjustable platform to support multiple configurations of the CMD-Tiny meter's control unit and probe position, relative to the robot's main body, and study the tradeoff between measurement consistency and robot traversability to identify an optimal sensor placement (see Section 3).

Figure 2.2 provides multiple views of the designed platform in CAD (Fusion 360), with all of the designed parts included in the final assembly. PVC tubes of 80 *cm* length, 10 *mm* diameter, and *SCH* 80 wall thickness serve as the basis for an expandable and rigid structure to mount the sensor (cylindrical part), as well as its associated data collection logger (rectangular part placed on top of the two long PVC tubes). Additionally, a shorter PVC tube of 30 *cm* length has been installed, along with an intermediate support mounted on the Jackal's front bumper, to enhance the stability of the overall EMI module. On the robot side, two supports are installed on the Jackal's top chassis to mount the two PCV tubes, and a unified support has been installed on the opposite side to hold the sensor's cylindrical probe. The latter support is crucial in allowing for reconfiguring the position and height from the ground of the probe and can be sturdily fixed in place when in use. Also, the total length of the sensor holder platform can be modified by the supports that are installed on the top chassis. Fabricated components of the platform were all 3D printed in polylactic acid (PLA) material.

2.3   DEVELOPMENT AND CALIBRATION FOR OPTIMAL SENSOR PLACEMENT

The optimal sensor placement in terms of the robot body is determined by two factors: 1) electromagnetic interference from the robot chassis and its onboard electronics, and 2) robot terrain traversability. The optimal solutions to each of these factors on their own are in fact opposing each other. Indeed, to minimize electromagnetic interference, the sensor should be placed as far from the robot as possible. On the contrary, the longer the extension of the sensor holding platform from the robot's center of mass, the worse it becomes to overcome obstacles without risking colliding the sensor with the ground (Figure 2.2), in addition to increasing the moment arm of the cantilever, which in turn would increase the required motor torque to move without tilting forward (that also leads to higher power consumption and lower operational time). In this section, we study optimal sensor placement in two distinct settings. First (Section 2.3.1), we benchmark experimentally the robot's interference on the EMI measurements for a given fixed distance from the ground ($d_h = 5$ *cm*) and for varying distances of the probe ($d_b \in [10, 100]$ *cm*). Second (Section 2.3.2), we determine terrain traversability while considering a subset of viable distances identified from the first step and two distinctive sensor height values to better understand the role of robot oscillations onto potential probe collisions with the ground. To make this process systematic, we create a realistic simulated environment and conduct this initial set of traversability evaluations in simulation. This process yields a set of candidate probe placement distances (d_b) as a function of probe height off the ground (d_h). Additionally, we perform feasibility experimental testing in measuring continuous ECa over small areas to both narrow down the range of viable configurations and to validate optimal sensor placement (Section 2.3.3).

2.3.1 DETERMINATION OF ROBOT INTERFERENCE IN SOIL CONDUCTIVITY MEASUREMENTS

The first step to be examined concerns the robot's electromagnetic interference in the soil conductivity measurements. Specifically, the robot is equipped with a high-torque 4×4 motored drivetrain and various onboard sensors such as a 3D LiDAR and the GNSS receiver. Given this setup, the robot generates electromagnetic fields that may interfere with the electromagnetic field generated by the GF CMD-Tiny sensor as well as the secondary current read by the sensor, and thus cause distorted measurements.

Experimental Procedure: To examine and mitigate the robot's interference on the EMI measurements, three independent field experiments were conducted under different robot-sensor configurations, in which the sensor was placed at predefined positions away from the robot's body to monitor the level of saturation. The fields that were selected for this purpose included a bare field, an olive tree grove, and an orange tree grove, which are located at the USDA-ARS U.S. Salinity Laboratory at the University of California, Riverside (33°58'21.936" N, −117°19'13.5732" E). In each field experiment, five distinct field points were selected to measure the soil conductivity. To get a broad spectrum of values for better understanding of the robot's electromagnetic interference, sampling points were either irrigated recently or non-irrigated. At the beginning of each experiment, handheld measurements were performed with the CMD-Tiny sensor, with no presence of any device that may generate additional electromagnetic fields. These measurements serve as the baseline. With these as a reference, we then placed the UGV with the CMD-Tiny sensor in different configurations and repeated the data collection process, each time increasing their relative distance within the range of [10, 100] *cm* at a step of 10 *cm*.

Soil conductivity data collected from these field tests are shown in Table 2.1. Column "∞" represents the handheld measurements where there is no appearance of external electromagnetic interference; the reported values are used as reference measurements. The columns starting from 10 *cm* until 100 *cm* represent the soil conductivity measurements that were made at the corresponding relative distances of the robot and the sensor's probe, at the same locations as in the handheld case. Mean and one-standard deviation values from the five trials in each of the 33 distinctive cases shown in Table 2.1 are also provided.

Key Findings: Results of this first experimental benchmark reveal that the optimal placement of the EMI sensor has a lower threshold of $d_b = 40$ *cm* away from the UGV body chassis. Specifically, numerical values reported in Table 2.1 show that the saturation is notable in the measurements when the sensor is less than 40 *cm* away from the robot's body. Shorter distances ({10, 20, 30} *cm*) exhibit both significantly higher mean values and more variability (i.e., higher one-standard deviation) across all three fields. Especially the $d_b = 10$ *cm* case exhibits excessive interference and clear evidence of saturation (the bare field depicts this most clearly). In

TABLE 2.1

Soil Conductivity Measurements in Evaluating Robot Electromagnetic Interference

Field Type	Distance between the robot and the probe (d_b) for constant probe height off the ground ($d_h = 5$ cm)										
mS/m	∞	10 cm	20 cm	30 cm	40 cm	50 cm	60 cm	70 cm	80 cm	90 cm	100 cm
	13.9	39.1	85	38	26.2	19.8	17.1	15.6	14.9	14.6	14.6
	9.9	36.8	58.4	28.8	18	13.4	11.6	10.8	10.4	10.2	10.1
Bare Field	10.9	−2	88.7	38.8	24.4	16.2	13.2	12.1	11.8	11.4	11.2
	13.2	−17.8	78.4	33.8	22.5	18.1	15	14.1	13.7	13.3	13.1
	10.3	66	61.8	30.3	20.2	15.4	12.8	11.5	11	10.7	10.5
Mean (μ)	11.64	24.42	74.46	33.94	22.26	16.58	13.94	12.82	12.36	12.04	11.90
Standard Deviation (σ)	1.80	33.83	13.67	4.47	3.26	2.46	2.15	1.98	1.89	1.85	1.90
	25.3	77.8	90.2	51.9	37.4	30.6	28	26.5	26	25.6	25.5
	26	47.2	96.1	51.3	36.3	30.5	28.1	27	26.6	26.4	26.2
Olive Tree Grove	21.8	7.8	106.3	50.6	34.1	27.8	24.4	23.1	22.4	22.1	21.9
	30.3	50.8	105.3	55.2	41.3	35.2	33.1	31.6	31	30.8	30.6
	23.2	59.4	87.9	48.1	32.3	29	25.6	24.5	24.1	23.8	23.6
Mean (μ)	25.32	48.60	97.16	51.42	36.28	30.62	27.84	26.54	26.02	25.74	25.56
Standard Deviation (σ)	3.25	25.69	8.44	2.56	3.43	2.81	3.34	3.23	3.24	3.28	3.28
	27.3	73.9	90.7	51.7	35.7	31.4	29.2	28.3	27.9	27.6	27.5
	23	69.2	85.2	47.3	33.6	27.7	25.3	24.2	23.6	23.4	23.2
Citrus Tree Grove	29.2	73.1	90.5	53.5	39	33.6	31.3	30.2	29.8	29.5	29.3
	32.2	35.6	101.7	57.8	43.1	37.2	34.7	33.4	32.8	32.5	32.3
	22.7	50.3	96.5	48.9	33.1	27.6	25.3	24.6	23.4	23.1	23
Mean (μ)	26.88	60.42	92.92	51.84	36.90	31.50	29.16	28.14	27.50	27.22	27.06
Standard Deviation (σ)	4.07	16.87	6.33	4.11	4.17	4.08	4.03	3.87	4.05	4.02	4.00

contrast, larger d_b values exhibit smaller interference, as it can be readily verified by the reported means and one-standard deviations that converge to the baseline values. In addition, it can be observed that after some threshold distance, means and one-standard deviations do not vary significantly. In this work, this upper threshold is selected at $d_b = 70$ cm.

Another interesting observation concerns the cross-field variability. The olive and citrus orchards that are regularly irrigated yield similar values across all tests, whereas the bare field (not irrigated) has lower ECa values (as expected). However, the measured values appear to converge faster (i.e., in shorter distances) in the irrigated fields compared to the bare field. We can associate this finding with the fact that in irrigated fields the reported ECa values are a function of both actual soil electrical conductivity and electromagnetic interference from the robot, and

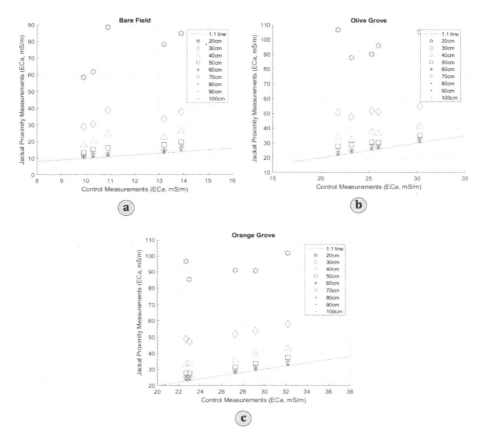

FIGURE 2.3 One-to-one control lines of soil conductivity measurements for the different robot-sensor distances at the three fields.

the former dominates more rapidly as the probe distance increases. In contrast, in the bare field, the actual ECa level attains much lower values, hence readings are more susceptible to electromagnetic interference and larger distances appear to be required to dampen down the interference's effect. This finding can be a useful tool to correlate soil salinity over the same field before and after planting, and while growing.

We can further justify the selection of the lower and upper thresholds for d_b by inspecting measurement linearity compared to the baseline (Figure 2.3) and via a Pearson correlation test (Figure 2.4) and subsequent linear regression (Figure 2.5). According to the 1:1 control lines in all three panels of Figure 2.3 that correspond to each tested field, the saturation is notable in the measurements when the sensor is less than 40 *cm* away from the robot's body; obtained values for 30 *cm* and below clearly deviate from the other cases, while obtained values of 70 *cm* are getting closely clustered together and approaching the 1:1 control line.[7] The case of $d_b = 40$

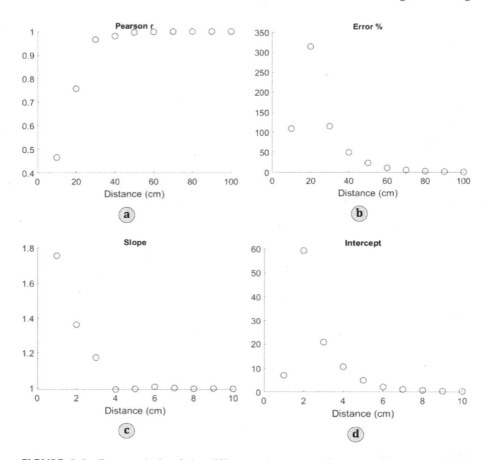

FIGURE 2.4 Data analysis of the different robot-sensor distances with respect to the handheld-collected (baseline) data: (a) Pearson correlation test; (b) error percentage in measurements; (c) the slope value; and (d) the intercept value are depicted.

cm appears to be the switching point. Although it could be argued based on Table 2.1 and Figure 2.3 that this case demonstrates high differentiation from the baseline (especially at the bare field), the Pearson correlation test (Figure 2.4a) yields a value of 0.98 for the distance of 40 *cm*, which approximates the value of 1 and thus can be considered as a constant linear offset. Slopes and linear regression results also provide additional supporting evidence for picking $d_b = 40$ *cm* as the lower threshold for further analysis. It can also be readily verified that there are no significant differences from setting the probe farther than 70 *cm* away from the robot body, as obtained values appear to have stabilized very close to the respective baseline

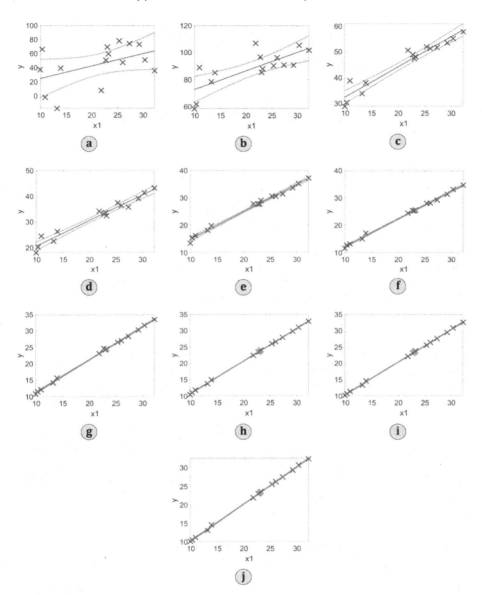

FIGURE 2.5 Linear regression graphs of the captured conductivity data with respect to the baseline values, for evaluating the robot interference in the EMI measurements. Panels (a)–(j) correspond to each of the cases in the range [10, 100] *cm* of robot-sensor distance, respectively. Upper and lower bounds of the fit are depicted with continuous curves (in red).

FIGURE 2.6 Instances of the simulated environment created in this work. (a) Final 3D mesh in the Agisoft Metashape software. (b) The simulated Jackal robot equipped with the designed sensor platform spawned in the simulated environment.

measurement, in all three fields. Considering that the farther we place the probe the worse its terrain traversability capacity (see next section too), we deduce that $d_b = 70$ *cm* is an appropriate upper threshold for further analysis. Finally, Figure 2.4d shows the intercept coefficient decrease as the probe distance increases, which corroborates all previous observations.

Given all these remarks, we validate that keeping the sensor at a certain distance and farther from the robot's body frame leads to a decrease of the included noise in the conductivity measurements because of electromagnetic interference and increases the linearity of the given readings with respect to the reference (handheld) measurements. As such, we select the fixed distances of $d_b \in \{40, 50, 60, 70\}$ *cm* to further evaluate the robot's traversability capacity and thus help select the optimal one.

2.3.2 EVALUATION OF ROBOT PLATFORM MANEUVERABILITY THOUGH GAZEBO SIMULATION

With the set of candidate sensor-to-body distance values having been identified ($d_b \in \{40, 50, 60, 70\}$ *cm*), the next step is to examine the robot's ability to move over various uneven terrain fields while carrying the probe without the probe colliding with the ground as the robot negotiates dips and bumps. In general, each platform configuration may have a distinct effect on the robot's traversability capacity, sensor stability, and thus sensor readings. In an effort to make this process scalable and generalizable, we develop a realistic simulation environment and test robot traversability and probe oscillations for different sensor placement configurations over varied sets of emulated terrains.

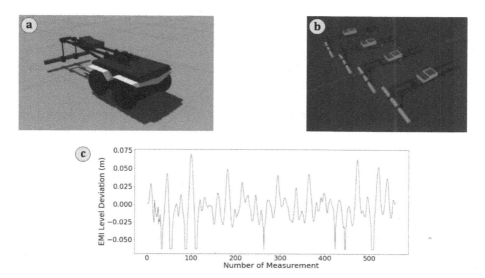

FIGURE 2.7 (a) Detailed view on the simulated robot with the sensor mounted. (b) Generated 3D models of the EMI platform in the configuration. (c) Time-series chart of level deviations of the sensor during its random roam in the simulated environment.

Experimental Procedure: We employed the Gazebo Robotics simulator [50] and considered three key steps: 1) generation of 3D model maps based on aerial imagery, 2) generation of the robot and sensor models, and 3) setup of the simulated experiments and measured variables. The first step allows for the generation of a realistic emulated environment based on real imagery data. We used the photogrammetric software Agisoft Metashape to create a 3D world model based on aerial imagery data that were captured just south of the Center for Environmental Research & Technology (CERT) at the University of California, Riverside (33°59'59.563267" N, −117°20'5.769141" E). By using 97 aerial photos obtained from different positions above the CERT field, we applied photo alignment and stitching through the Agisoft Metashape libraries and generated a colored pointcloud of 25,904 points. In addition, by enhancing the sparse pointcloud with more correspondences from the stitching process, a dense pointcloud was generated containing ≈ 9 million points; the latter was then transformed into a 3D mesh model by applying point interpolation. Figure 2.6a shows the final form of the 3D mesh of the generated world, inside the Agisoft Metashape interface, and Figure 2.6b depicts the integration of the world into the Gazebo simulator along with the simulated real-scale robot.

A Gazebo model for the Clearpath Robotics Jackal robot employed herein is already publicly available.[8] We developed a custom Gazebo model for the GF CMD-Tiny sensor and integrated it into the main robot model. Figure 2.7 depicts these models. Four separate Gazebo models of the sensor were generated (Figure 2.7b), each corresponding to one of the considered probe distances $d_b \in \{40, 50, 60, 70\}$ *cm*. The sensor's height off the ground was adjusted to be $d_h \in \{6, 11\}$ *cm*, which approximates the average height of sensor placement during handheld continuous measurements. The simulated Jackal model weighs 17 *kg*, similar to the real one, and the sensor platform with the onboard CMD-Tiny sensor is at 4 *kg*. To evaluate the GF CMD-Tiny sensor oscillations in the z-axis during the Jackal's movement, a ranging plugin was also developed in Gazebo, to provide continuous information about the distance of the EMI sensor from the ground. The Gazebo plugin publishes continuously the ranges of the sensor body to the ground through a *sensor_msgs/ LaserScan* ROS topic, thus allowing us to capture the deviation data in real time (one such example is shown in Figure 2.7c). Note that the simulated setup features the same ROS libraries as the real robot does, and it can be teleoperated or commanded to move to a goal pose (i.e., position and orientation) in the local frame, as it has an integrated RTK-GNSS antenna as well.

To evaluate the rigidity and robustness of the designed platform, we conducted three independent simulated experiments inside the emulated field in the Gazebo world for the selected platform configurations (parameters d_b and d_h). In these experiments, the Jackal is commanded to follow specific trajectories of different difficulty terrain types, namely a straight path in a slightly planar area, a straight path in a rockier area, and a complex path with mixed type terrain (Figure 2.8). During the trajectory following, the robot measures the vertical oscillations of the EMI sensor with respect to the ground. Each individual case is repeated for five times, thus giving rise to a total of 120 (simulated) experimental trials in this specific benchmark.

Key Findings: Results of this second experimental benchmark are contained in Table 2.2. First, we confirm (as anticipated) that the increase of the sensor distance

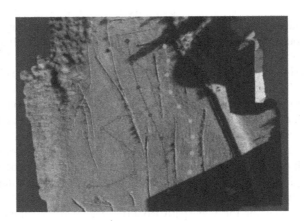

FIGURE 2.8 Planned trajectories for testing robot traversability in the simulated Gazebo world. The blue (6-node) trajectory lies on a smooth region of the map, the green (8-node) is on an uneven/rocky land area of the map, and the red (13-node) contains a longer trajectory over mixed type terrain (best viewed in color).

TABLE 2.2

Measured Sensor Oscillations during Simulated Surveys in the Gazebo Environment

	Smooth Land			Rocky Land			Mixed-Terrain Land		
$d_b - d_h$	mean (cm)	σ (cm)	variance (cm²)	mean (m)	σ (cm)	variance (cm²)	mean (cm)	σ (cm)	variance (cm²)
40cm – 6cm	−0.50	2.57	6.61e-02	−0.51	3.15	9.94e-02	−0.54	2.58	6.65e-02
50cm – 6cm	−0.60	3.13	9.81e-02	−0.52	3.13	9.77e-02	−0.68	2.80	7.83e-02
60cm – 6cm	−0.83	2.97	8.81e-02	−0.49	3.24	0.10	−0.79	3.02	9.10e-02
70cm – 6cm	0.68	4.11	0.17	−0.61	3.91	0.15	−0.71	3.20	0.10
40cm – 11cm	−0.84	2.97	8.83e-02	−0.78	3.37	0.11	−1.03	3.35	0.11
50cm – 11cm	−1.12	3.39	0.11	−1.03	4.07	0.17	−1.25	3.66	0.13
60cm – 11cm	−1.47	4.37	0.19	−0.93	3.94	0.16	−1.57	4.47	0.20
70cm – 11cm	−1.23	3.69	0.14	−1.62	4.76	0.23	−1.50	4.04	0.16

relative to the robot body makes the robot less stable and in turn leads to increased probe oscillations when traversing uneven terrain. However, it turns out that the cases of $d_h = \{50, 60\}$ *cm* lead to very similar platform oscillations in terms of reported variance, which in fact are close to the shortest case of 40 *cm* and more steady compared to the longest 70 *cm* case. In more detail, and with reference to Table 2.2, the increase of the distance in the probe placement results in increased mean and variance of position deviations while moving, for both tested sensor height levels off the ground. Mounting the sensor closer to the robot's body and keeping a shorter robot-sensor footprint, the robot is more stable in either smoother or rougher terrain types. For the $d_h = 6$ *cm* case, placing the sensor at $d_b = \{50, 60\}$ *cm* achieves similar results by keeping ≈ 1 *cm* of difference in standard deviation of the 70 *cm* case. Also, even though 40 *cm* is the less shaky solution overall, the cases of $d_b = \{50, 60\}$ *cm* have less than ≈ 0.5 *cm* difference in standard deviation. Additionally, by comparing across experiments, the 50 *cm* case appears to score less than 10^{-4} m^2 of variance. These observations are consistent also in the case of $d_h = 11$ *cm*, by having the smallest sensor displacement when $d_b = 40$ *cm* and with the cases of $d_b = \{50, 60\}$ *cm* performing equivalently. Based on these observations, we deduce that the cases of setting $d_b = \{50, 60\}$ *cm* may offer the best tradeoff in terms of noise and steadiness compared with the shorter 40 *cm* case, in which the probe is placed close to the robot's chassis and is affected more by the electromagnetic interference, as shown in Section 2.3.

A second observation from these results concerns the sensor's height off the ground, d_h. Inspection of the results for different sensor height values d_h in Table 2.2, we notice a similar oscillating behavior, whereby a shorter height (i.e., $d_h = 6$ *cm*) might be preferred to reduce variations from the ground. This becomes even more critical considering that, during real surveys, it is hard for a farm worker who uses the handheld sensor to keep the sensor consistently at a stable height off the ground while walking. For these reasons we chose to use $d_h = 6$ *cm*.

2.3.3 Preliminary Feasibility Experimental Testing of Boundary Configurations

The last set of calibration tests for optimal sensor placement included validation of the preliminary feasibility of the robot-sensor setup to measure ECa continuously. Based on the aforementioned results, we elected to study in these validation experiments the two boundary cases of probe distance (i.e., $d_b = \{40, 70\}$ *cm*), at a constant height of $d_h = 6$ *cm*.[9] By doing so, we can get a better understanding of the effects of placing the sensor closer or further away from the robot on *continuous* ECa measurements.

Experimental Procedure: The validation experiments took place in the same field that was emulated for the aforementioned simulation testing. Without loss of generality, we considered two distinct cases: 1) a straight-line trajectory with $d_b = 70$ *cm* and $d_h = 6$ *cm*, and 2) a U-shaped trajectory with $d_b = 40$ *cm* and $d_h = 6$ *cm*. In each case we conducted three independent trials. The starting and end positions as well as intermediate waypoints were the same for each set of trials. The physical setup, experimental field, and the two types of trajectories are depicted in Figure 2.9. Manual data collections with the handheld sensor (three for each

FIGURE 2.9 (a) The Jackal robot equipped with the platform holding the CMD-Tiny instrument at the configuration of $d_b = 50$ *cm* and $d_h = 6$ *cm*. (b) Instance of robot during the soil ECa data collection for validation. (c) Validation testing considered two distinctive trajectories, one following a straight line and another performing a U-shaped curve.

trajectory type) were also performed to serve as the baseline. In total, we conducted 12 experimental trials for validation in the bare field (six robotized and six manual). Obtained soil ECa measurements were georeferenced via RTK-GNSS in the robotized measurement and the sensor's embedded GPS in the manual measurements. Collected soil ECa data were used to generate a custom raster of the surveyed area; then we used kriging interpolation through exponential semivariogram to obtain the ECa map, which was embedded onto satellite imagery via the ESRI ArcGIS 10.8.2 software.

Key Findings: Obtained results visualized based on aggregated soil conductivity plots and the corresponding ECa maps for the cases of $d_b = 70$ *cm* (straight line trajectory) and $d_b = 40$ *cm* (U-shaped trajectory) are depicted in Figure 2.10 and Figure 2.11, respectively. Foremost, results validate that the shorter probe distance placement demonstrates a higher difference on average sensor readings compared to the longer distance setting against the manual baseline data (case $d_b = 40$ *cm*, robotized: $\{\mu = 13.32, \sigma = 0.99\}$ *mS/m* and manual: $\{\mu = 7.55, \sigma = 0.92\}$ *mS/m*; case $d_b = 70$ *cm*, robotized: $\{\mu = 11.40, \sigma = 1.83\}$ *mS/m* and manual: $\{\mu = 9.34, \sigma = 1.59\}$ *mS/m*). Despite this increased difference, the standard deviations are close between robotized and manual measurements in both cases. This indicates that there is consistency among the two types of measurements and that the observed offsets in robotized measurements compared to their respective manual baselines can in fact be treated as constant offsets. Further evidence in support of the constant offset presence can be obtained by the Pearson Correlation Coefficient (PCC). In both cases (straight line and U-shape), the robotized ECa measurements showcase high linearity with their manual counterparts. Specifically, by removing outlier measurements that lie out of the $\pm 2\sigma$ measurement distribution, the straight-line trajectory attains a PCC of 0.95, whereas the U-shaped trajectory reaches a PCC of 0.88 (despite significant electromagnetic interference as demonstrated in static tests discussed in Section 2.3.1), compared with manual measurements that were conducted on exactly the same paths. These findings can be visually corroborated by the obtained ECa maps in panels (b) and (c) of Figure 2.10 and Figure 2.11, with the main difference being that the color differential in the shorter probe placement case of $d_b = 40$ *cm* (Figure 2.11) being more pronounced due to the larger constant offset in measurements compared to the longer probe placement case of $d_b = 70$ *cm* (Figure 2.10).

In all, preliminary feasibility testing validates we can perform trustworthy continuous ECa measurements with the developed robotic setup even at the two boundary configurations. Results verify the system's high linearity with respect to manually collected (handheld) data, and that the constant additive offset on the overall ECa measurements caused by the robot's electromagnetic interference does not significantly affect the linearity of obtained measurements in the end. Putting everything together, we conclude that the configuration $\{d_b = 60$ *cm*, $d_h = 6$ *cm*$\}$ can serve as an appropriate tradeoff that combines lower robot-to-sensor signal interference and higher maneuverability, and it is thus selected as the optimal configuration to conduct field-scale experiments. These are discussed next.

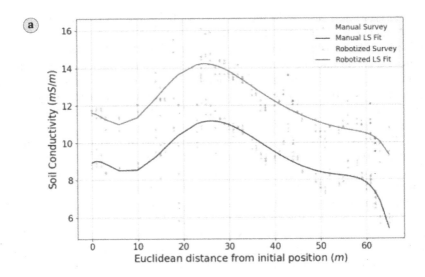

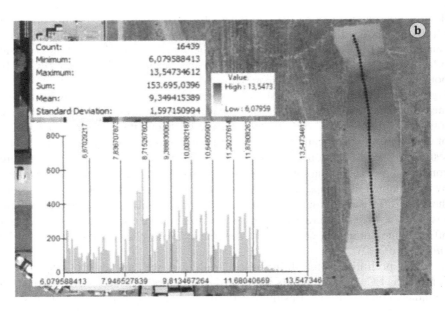

FIGURE 2.10 (a) The soil conductivity curves of the straight-line trajectory case in the bare field. Fitted graphs of both measurement curves correspond to an 8th grade least squares fit. (b)-(c) Soil ECa maps corresponding to manually collected (handheld) and robotized-collected data, respectively, computed by applying kriging interpolation through exponential semivariogram. Each panel also contains the value-based color scale, the map statistics, and the histogram of the conductivity values.

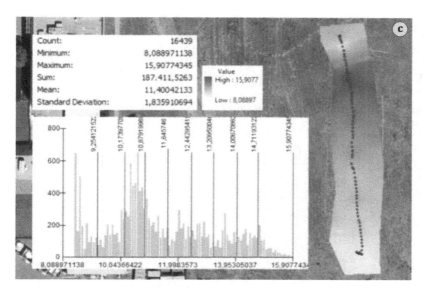

FIGURE 2.10   (Continued)

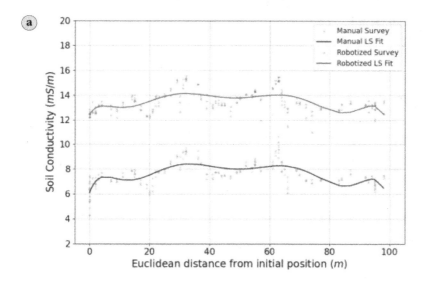

FIGURE 2.11   (a) The soil conductivity curves of the U-shaped trajectory case in the bare field. Fitted graphs of both measurement curves correspond to an 8th grade least squares fit. (b)-(c) Soil ECa maps corresponding to manually-collected (handheld) and robotized-collected data, respectively, computed by applying kriging interpolation through exponential semivariogram. Each panel also contains the value-based color scale, the map statistics, and the histogram of the conductivity values.

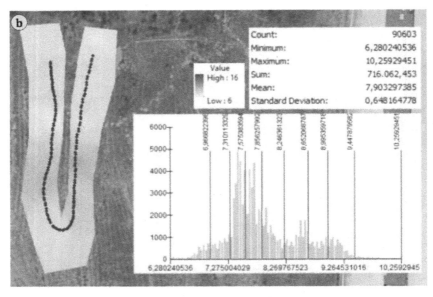

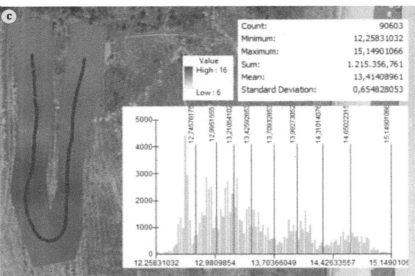

FIGURE 2.11 (Continued)

2.4 FIELD-SCALE EXPERIMENTS

The analysis conducted so far has helped determine an optimal sensor setup ($\{d_b = 60\ cm,\ d_h = 6\ cm\}$) for the robot considered in this study that balances between electromagnetic interference caused by the robot and its electronic components and robot traversability capacity of uneven terrain. The analysis has also helped validate the preliminary feasibility of collecting continuous soil ECa measurements over small bare-field areas, with obtained results being trustworthy and consistent to manually collected baselines. We now turn our attention to robot-assisted continuous soil ECa measurements over larger fields.

Experimental Procedure: We performed continuous soil ECa measurements in two distinctive fields, an olive tree grove (not irrigated recently with respect to data collections) and a citrus tree grove (irrigated prior to data collections). The arid canopy of olive trees is located close to the USDA-ARS U.S. Salinity Laboratory at the University of California, Riverside (UCR) (33°58'21.936" N, −117°19'13.5732" E), whereas the freshly irrigated citrus orchard is located within the UCR Agricultural Experimental Station fields (AES; 333°57'52.0272" N, −117°20'13.7184" E). It is worth noting that the latter case in fact contained various soil conditions (highly irrigated/wet parts and arid parts), and hence helped evaluate 1) the proposed robot's performance in the same survey as terrain conditions vary, and 2) if the robot-assisted soil conductivity curves and soil ECa maps match those corresponding to manual data collections. Experimental setups and snapshots of the two fields are depicted in Figure 2.12. For the olive tree grove, we considered an area of two full tree rows covered following a U-shaped trajectory, whereas for the citrus tree grove, we considered an area of roughly three tree rows covered following an S-shaped trajectory. Manual data collections with the handheld sensor (three in each field) were also performed to serve as the baseline. In total, we conducted 12 experimental trials for overall system field-scale evaluation in the olive and citrus tree groves (six robotized and six manual). Obtained soil ECa measurements were georeferenced via RTK-GNSS in the robotized measurement and the sensor's embedded GPS in the manual measurements. Collected soil ECa data were used to generate a custom raster of the surveyed area; then kriging interpolation through exponential semivariogram helped obtain the ECa map, which was embedded onto satellite imagery via the ESRI ArcGIS 10.8.2 software.

Key Findings: Results from field-scale experiments for olive and citrus tree groves are shown in Figures 2.13–2.16. Collected raw data from each considered case, aggregated soil conductivity data comparing robotized to manual baselines, as well as the corresponding ECa maps are presented. It can be readily verified from the graphs that robot-assisted continuous ECa measurements approximate very well the manually collected baselines in both cases.

For the case of the olive tree grove (arid field), raw measurement plots (Figure 2.13a and Figure 2.13b) match either very well. The visual observation is corroborated via the mean measurement plots (Figure 2.13c), where PCC reaches a score of 0.97. The mean conductivity value in the robotized case is 14.23 *mS/m*, compared to the 10.56 *mS/m* mean value of the manual case, which represents a fixed increase that can be seen by the curvature of the polynomial fits. Additionally, the standard deviations are closely matching robotized: 1.65 *mS/m*; manual: 1.75 *mS/m*), which demonstrates consistency across both soil ECa measurement means. Output ECa

FIGURE 2.12 Field-scale data collections instances: (a) robotized in the olive tree grove and (b)–(c) manual/robotized in the orange tree grove.

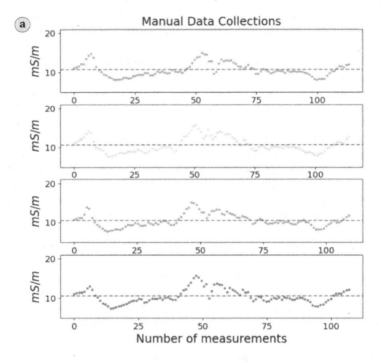

FIGURE 2.13 Olive tree grove case. (a) and (b) Graphs of the raw conductivity measurements by using directly the sensor and via the robot, respectively. (c) Soil conductivity plots of both handheld and robot cases. Fitted plot of both measurement curves is an 8th grade least squares fit.

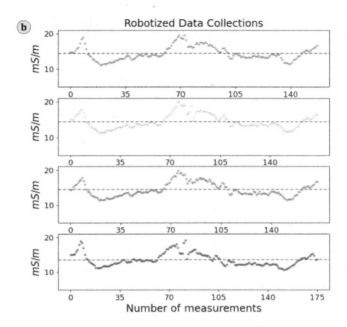

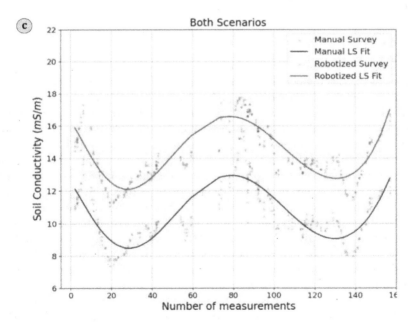

FIGURE 2.13   (Continued)

maps (Figure 2.14) are also very similar, with a PCC value of 0.97 in a pixel-wise correlation. It is notably clear that the robotized results are close to the manual ones, despite the constant positive offset on the level of the measured soil apparent electrical conductivity.

Similar observations can be readily made for the case of the citrus tree grove. Recall that this field was recently irrigated prior to data collections, hence there was more terrain variability; this ranged from muddy soil to normally irrigated areas and more dry parts of the field areas. From a robot operation standpoint, the proposed system demonstrates robust behavior even in this diverse and more demanding survey case, and it is noteworthy that the robot's traversability in the irrigated turf was efficient even when navigating over mud. Raw data graphs (Figure 2.15a and Figure 2.15b) demonstrate several notable peaks in soil conductivity measurements that were caused by traversability in the muddy terrain and the well-irrigated soil. These peaks were captured in both manual and robot-assisted surveys. The graphs shown in Figure 2.15c have a PCC of 0.90, which demonstrates the proposed robot's robustness in even muddy and quite diverse soil-ECa-level fields. The polynomial fits of both plots (Figure 2.15c) present similar curvature, with an additive offset increase in the robot's case, caused by the electromagnetic interference, whereas the pixel-wise correlation of the output ECa maps reaches a value of 0.96 (Figure 2.16). Visual inspection of the obtained soil ECa maps also supports the aforementioned findings.

2.5 CONCLUSION

In this study, we presented a robotized means to perform precise and continuous soil apparent electrical conductivity (ECa) measurement surveys. Our proposed solution involves a ground mobile robot equipped with a customized and adjustable platform to hold an electromagnetic induction (EMI) sensor. The optimal placement of the EMI sensor is determined by satisfying two competing objectives: 1) the minimization of static electromagnetic interference in measurements and 2) the facilitation of robot traversability in the field. Extensive experimental evaluation across static calibration tests and over different types of fields concludes as to the optimal EMI configuration setup to be used for large-field testing. Throughout a series of real field experiments, our study demonstrates that the obtained robot-assisted soil conductivity measurements present high linearity compared to the ground truth (data collected manually by a handheld EMI sensor) by scoring more than 90% in Pearson correlation coefficient in both plot measurements and estimated ECa maps generated by kriging interpolation. The proposed platform can deliver high-linearity scores in real survey scenarios, at an olive and citrus grove under different irrigation levels, and serve as a robust tool for large-scale ECa mapping in the field, with the potential development of fully autonomous behavior. Future work will focus on integration within a task and motion framework for informative proximal sampling, as well as integration with physical sampling means to perform multiple tasks simultaneously.

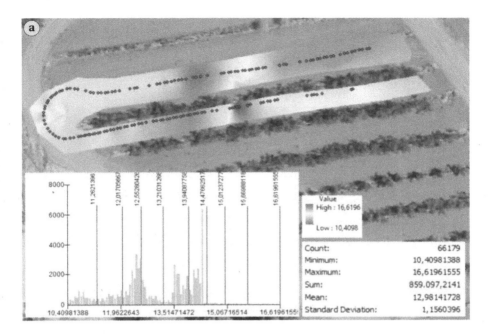

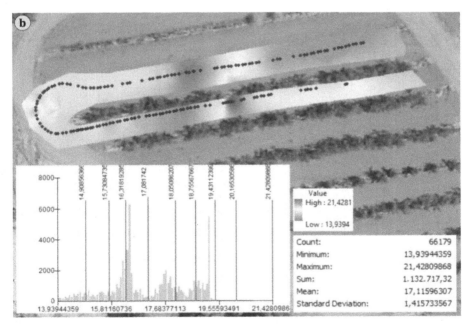

FIGURE 2.14 Obtained soil ECa maps in the olive tree grove for (a) the manual and (b) robotized cases. Maps have been created by applying kriging interpolation through exponential semivariogram. Each map also depicts the value-based color scale, the map statistics, and the histogram of the conductivity values.

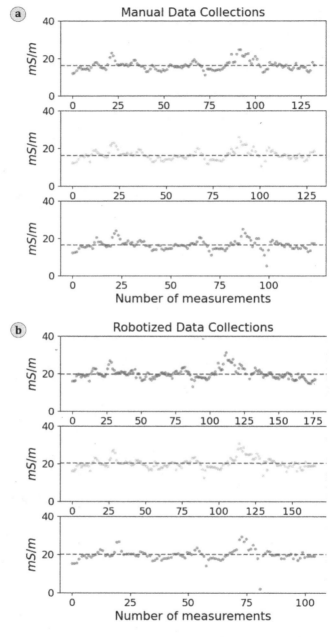

FIGURE 2.15 Citrus tree grove case. (a) and (b) Graphs of the raw conductivity measurements by using directly the sensor and via the robot, respectively. (c) Soil conductivity plots of both handheld and robot cases. The color-shaded areas indicate the filled areas of soil conductivity values of the corresponding manual and robotized measurements. Fitted plot of both measurement curves is an 8th grade least squares fit.

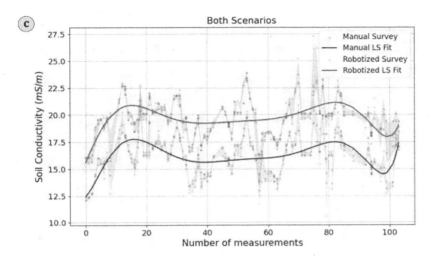

FIGURE 2.15 (Continued)

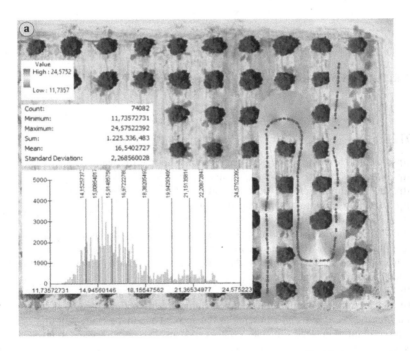

FIGURE 2.16 Obtained soil ECa maps in the citrus tree grove for (a) the manual and (b) robotized cases. Maps have been created by applying kriging interpolation through exponential semivariogram. Each map also depicts the value-based color scale, the map statistics, and the histogram of the conductivity values.

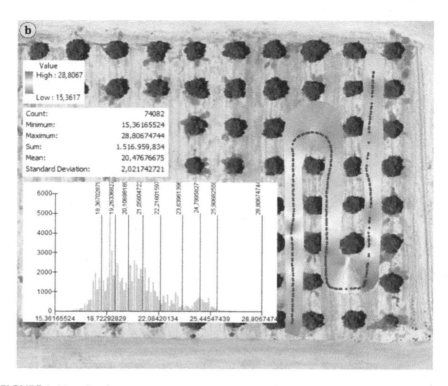

FIGURE 2.16 (Continued)

NOTES

1 https://clearpathrobotics.com/jackal-small-unmanned-ground-vehicle/
2 https://clearpathrobotics.com/husky-unmanned-ground-vehicle-robot/
3 https://clearpathrobotics.com/warthog-unmanned-ground-vehicle-robot/
4 https://farm-ng.com/products/la-maquina-amiga
5 http://wiki.ros.org/mavros
6 http://docs.ros.org/en/jade/api/robot localization/html/navsat transform node.html
7 Note that the case of 10 *cm* has significant interference and widely varied values (at cases even negative as shown in Table 2.1) and is henceforth not shown in these graphs to improve visual clarity of the figure.
8 http://wiki.ros.org/Robots/Jackal
9 We wish to highlight at this point that the obtained results so far suggest that the configuration $\{d_b = 60\ cm,\ d_h = 6\ cm\}$ can serve as the optimal one in the context of this work. This is the configuration we test in field-level experiments in Section 2.4 that follows. However, for completeness purposes, and in an effort to better explain the effects of placing the sensor closer or further away from the robot on soil ECa measurements, we tested also the two boundary configurations (which are still viable in principle) but at a smaller-scale experimental setup compared to the field-level experimental setups in Section 2.4.

REFERENCES

[1] R. B. Daniels, "Geology of soils: Their evolution, classification, and uses," *Soil Science Society of America Journal*, vol. 37, no. 1, 1973.

[2] P. W. Birkeland, *Soils and Geomorphology*. Oxford: Oxford University Press, 1984.

[3] P. J. Greminger, Y. K. Sud, and D. R. Nielsen, "Spatial variability of field-measured soil-water characteristics," *Soil Science Society of America Journal*, vol. 49, no. 5, pp. 1075–1082, 1985.

[4] K. K. Tanji, *Salinity in the Soil Environment*, pp. 21–51. Dordrecht: Springer Netherlands, 2002.

[5] D. L. Corwin and E. Scudiero, "Field-scale apparent soil electrical conductivity," *Soil Science Society of America Journal*, vol. 84, no. 5, pp. 1405–1441, 2020.

[6] J. Huang, E. Scudiero, W. Clary, D. Corwin, and J. Triantafilis, "Time-lapse monitoring of soil water content using electromagnetic conductivity imaging," *Soil Use and Management*, vol. 33, no. 2, pp. 191–204, 2017.

[7] S. P. Friedman, "Soil properties influencing apparent electrical conductivity: A review," *Computers and Electronics in Agriculture*, vol. 46, no. 1, pp. 45–70, 2005.

[8] D. L. Corwin and S. M. Lesch, "Application of soil electrical conductivity to precision agriculture," *Agronomy Journal*, vol. 95, no. 3, pp. 455–471, 2003.

[9] V. Adamchuk, W. Ji, R. V. Rossel, R. Gebbers, and N. Tremblay, "Proximal soil and plant sensing," *Precision Agriculture Basics*, pp. 119–140, 2018.

[10] K. Noborio, "Measurement of soil water content and electrical conductivity by time domain reflectometry: A review," *Computers and Electronics in Agriculture*, vol. 31, no. 3, pp. 213–237, 2001.

[11] W. Li, H. M. Wainwright, Q. Yan, H. Zhou, B. Dafflon, Y. Wu, R. Versteeg, and D. M. Tartakovsky, "Estimation of evapotranspiration rates and root water uptake profiles from soil moisture sensor array data," *Water Resources Research*, vol. 57, no. 11, p. e2021WR030747, 2021.

[12] M. Siddiqui, V. Palaparthy, H. Kalita, M. Shojaei Baghini, and M. Aslam, "Graphene oxide array for in-depth soil moisture sensing toward optimized irrigation," *ACS Applied Electronic Materials*, vol. 2, no. 12, 2020.

[13] C. Peng, S. Vougioukas, D. Slaughter, Z. Fei, and R. Arikapudi, "A strawberry harvest-aiding system with crop-transport collaborative robots: Design, development, and field evaluation," *Journal of Field Robotics*, vol. 39, no. 8, pp. 1231–1257, 2022.

[14] M. Karthikeyan and D. Manimegalai, "Automatic soil moisture control and seed sowing using agrobot," in *2nd International Conference on Advance Computing and Innovative Technologies in Engineering (ICACITE)*, Greater Noida, India, pp. 375–379, 2022. https://doi.org/10.1109/ICACITE53722.2022.9823443

[15] D. An, H. Niu, L. Williams, and Y. Chen, "Smart Bi-eBikes (SBB): A low cost UGV solution for precision agriculture applications," *Proc. SPIE, Autonomous Air and Ground Sensing Systems for Agricultural Optimization and Phenotyping VII* 121140C (3 June 2022); https://doi.org/10.1117/12.2618728. Proceedings Volume 12114, Autonomous Air and Ground Sensing Systems for Agricultural Optimization and Phenotyping VII; 121140C (2022). https://doi.org/10.1117/12.2618728. Event: SPIE Defense + Commercial Sensing, 2022, Orlando, Florida, United States.

[16] R. Tazzari, D. Mengoli, and L. Marconi, "Design concept and modelling of a tracked UGV for orchard precision agriculture," in *2020 IEEE International Workshop on Metrology for Agriculture and Forestry (MetroAgriFor)*, Trento, Italy, pp. 207–212, 2020. https://doi.org/10.1109/MetroAgriFor50201.2020.9277577

[17] F. Sarghini, V. Visacki, A. Sedlar, M. Crimaldi, V. Cristiano, and A. D. Vivo, "First measurements of spray deposition obtained from UAV spray application technique," in *2019 IEEE International Workshop on Metrology for Agriculture and Forestry (MetroAgriFor)*, Portici, Italy, pp. 58–61, 2019. https://doi.org/10.1109/MetroAgriFor.2019.8909233

[18] S. Khan, M. Tufail, M. T. Khan, Z. A. Khan, J. Iqbal, and A. Wasim, "A novel framework for multiple ground target detection, recognition and inspection in precision agriculture applications using a uav," *Unmanned Systems*, vol. 10, no. 1, pp. 45–56, 2022.

[19] V. P. Subba Rao and G. S. Rao, "Design and modelling of an affordable UAV based pesticide sprayer in agriculture applications," in *2019 Fifth International Conference on Electrical Energy Systems (ICEES)*, Chennai, India, pp. 1–4, 2019. https://doi.org/10.1109/ICEES.2019.8719237

[20] P. Katsigiannis, L. Misopolinos, V. Liakopoulos, T. K. Alexandridis, and G. Zalidis, "An autonomous multi-sensor UAV system for reduced-input precision agriculture applications," in *2016 24th Mediterranean Conference on Control and Automation (MED)*, Athens, Greece, pp. 60–64, 2016. https://doi.org/10.1109/MED.2016.7535938

[21] J. Pak, J. Kim, Y. Park, and H. I. Son, "Field evaluation of path-planning algorithms for autonomous mobile robot in smart farms," *IEEE Access*, vol. 10, pp. 60253–60266, 2022.

[22] P. Tokekar, J. V. Hook, D. Mulla, and V. Isler, "Sensor planning for a symbiotic uav and ugv system for precision agriculture," *IEEE Transactions on Robotics*, vol. 32, no. 6, pp. 1498–1511, 2016.

[23] K. Dan, C. Mcleod, and A. Thomasson, "Improved autonomous mobile ground control point robot collaborates with UAV to improve accuracy of agriculture remote sensing," in *Proc. SPIE, Autonomous Air and Ground Sensing Systems for Agricultural Optimization and Phenotyping VII* (J. A. Thomasson and A. F. Torres-Rua, eds.), vol. 12114, p. 1211408. Orlando, FL: International Society for Optics and Photonics, SPIE, 2022. https://doi.org/10.1117/12.2622615

[24] C. Potena, R. Khanna, J. Nieto, R. Siegwart, D. Nardi, and A. Pretto, "Agricolmap: Aerial-ground collaborative 3d mapping for precision farming," *IEEE Robotics and Automation Letters*, vol. 4, no. 2, pp. 1085–1092, 2019.

[25] M. Schirrmann, R. Gebbers, E. Kramer, and J. Seidel, "Soil ph mapping with an on-the-go sensor," *Sensors*, vol. 11, no. 1, pp. 573–598, 2011.

[26] M. Polic, M. Car, J. Tabak, and M. Orsag, "Robotic irrigation water management: Estimating soil moisture content by feel and appearance," *CoRR*, vol. abs/2201.07653, 2022.

[27] X. Yu, C. Chang, J. Song, Y. Zhuge, and A. Wang, "Precise monitoring of soil salinity in China's yellow river delta using uav-borne multispectral imagery and a soil salinity retrieval index," *Sensors*, vol. 22, no. 2, 2022.

[28] T. C. Thayer, S. Vougioukas, K. Goldberg, and S. Carpin, "Routing algorithms for robot assisted precision irrigation," in *2018 IEEE International Conference on Robotics and Automation (ICRA)*, Brisbane, QLD, Australia, pp. 2221–2228, 2018. https://doi.org/10.1109/ICRA.2018.8461242

[29] J. Pulido Fentanes, A. Badiee, T. Duckett, J. Evans, S. Pearson, and G. Cielniak, "Kriging-based robotic exploration for soil moisture mapping using a cosmic-ray sensor," *Journal of Field Robotics*, vol. 37, no. 1, pp. 122–136, 2020.

[30] D. Tseng, D. Wang, C. Chen, L. Miller, W. Song, J. Viers, S. Vougioukas, S. Carpin, J. A. Ojea, and K. Goldberg, "Towards automating precision irrigation: Deep learning to infer local soil moisture conditions from synthetic aerial agricultural images," in *2018 IEEE 14th International Conference on Automation Science and Engineering (CASE)*, Munich, Germany, pp. 284–291, 2018. https://doi.org/10.1109/COASE.2018.8560431

[31] A. Lukowska, P. Tomaszuk, K. Dzierzek, and L. Magnuszewski, "Soil sampling mobile platform for agriculture 4.0," in *2019 20th International Carpathian Control*

Conference (ICCC), Krakow-Wieliczka, Poland, pp. 1–4, 2019. https://doi.org/10.1109/CarpathianCC.2019.8765937

[32] B. V. Dimaya, K. J. U. Kasilag, F. I. D. Ong, B. P. B. Ramirez, K. E. V. Ramirez, C. G. Pascion, N. M. Arago, and M. V. Padilla, "Mobile soil robot collector via smartphone with global positioning system for navigation," in *2018 IEEE International Conference on Humanoid, Nanotechnology, Information Technology, Communication and Control, Environment and Management (HNICEM)*, Baguio City, Philippines, pp. 1–6, 2018. https://doi.org/10.1109/HNICEM.2018.8666240

[33] D. Bourgeois, A. G. Bourgeois, and A. Ashok, "Demo: RoSS: A low-cost portable mobile robot for soil health sensing," in *2022 14th International Conference on COMmunication Systems & NETworkS (COMSNETS)*, Bangalore, India, pp. 436–437, 2022. https://doi.org/10.1109/COMSNETS53615.2022.9668355

[34] M. Campbell, K. Ye, E. Scudiero, and K. Karydis, "A portable agricultural robot for continuous apparent soil electrical conductivity measurements to improve irrigation practices," in *2021 IEEE 17th International Conference on Automation Science and Engineering (CASE)*, Lyon, France, pp. 2228–2234, 2021. https://doi.org/10.1109/CASE49439.2021.9551401

[35] H. Gharakhani, J. A. Thomasson, P. Nematzadeh, P. K. Yadav, and S. Hague, "Using under-canopy cotton imagery for cotton variety classification," in *Proc. SPIE, Autonomous Air and Ground Sensing Systems for Agricultural Optimization and Phenotyping VII* (J. A. Thomasson and A. F. Torres-Rua, eds.), vol. 12114, p. 121140B. International Society for Optics and Photonics, SPIE, 2022. https://doi.org/10.1117/12.2623034

[36] Y. Li, K.-D. Nguyen, and H. Dankowicz, "A robust adaptive controller for a seed refilling system on a moving platform," *IFAC-PapersOnLine*, vol. 49, no. 16, pp. 341–346. 5th IFAC Conference on Sensing, Control and Automation Technologies for Agriculture AGRICONTROL 2016.

[37] T. Guan, D. Kothandaraman, R. Chandra, A. J. Sathyamoorthy, K. Weerakoon, and D. Manocha, "Ga-nav: Efficient terrain segmentation for robot navigation in unstructured outdoor environments," *IEEE Robotics and Automation Letters*, vol. 7, no. 3, pp. 8138–8145, 2022.

[38] I. D. Miller, F. Cladera, T. Smith, C. J. Taylor, and V. Kumar, "Stronger together: Air-ground robotic collaboration using semantics," *IEEE Robotics and Automation Letters*, vol. 7, no. 4, pp. 9643–9650, 2022.

[39] B. P. L. Lau, B. J. Y. Ong, L. K. Y. Loh, R. Liu, C. Yuen, G. S. Soh, and U.-X. Tan, "Multi-agv's temporal memory-based rrt exploration in unknown environment," *IEEE Robotics and Automation Letters*, vol. 7, no. 4, pp. 9256–9263, 2022.

[40] Y. F. Chen, M. Everett, M. Liu, and J. P. How, "Socially aware motion planning with deep reinforcement learning," in *2017 IEEE/RSJ International Conference on Intelligent Robots and Systems (IROS)*, Vancouver, BC, Canada, pp. 1343–1350, 2017. https://doi.org/10.1109/IROS.2017.8202312

[41] V. Narayanan, B. M. Manoghar, V. Sashank Dorbala, D. Manocha, and A. Bera, "ProxEmo: Gait-based emotion learning and multi-view proxemic fusion for socially-aware robot navigation," in *2020 IEEE/RSJ International Conference on Intelligent Robots and Systems (IROS)*, Las Vegas, NV, pp. 8200–8207, 2020. https://doi.org/10.1109/IROS45743.2020.9340710

[42] H. Karnan, G. Warnell, X. Xiao, and P. Stone, "VOILA: Visual-observation-only imitation learning for autonomous navigation," in *2022 International Conference on Robotics and Automation (ICRA)*, Philadelphia, PA, pp. 2497–2503, 2022. https://doi.org/10.1109/ICRA46639.2022.9812316

[43] Z. Wang, X. Xiao, G. Warnell, and P. Stone, "Apple: Adaptive planner parameter learning from evaluative feedback," *IEEE Robotics and Automation Letters*, vol. 6, no. 4, pp. 7744–7749, 2021.

[44] T. Moore and D. Stouch, "A generalized extended kalman filter implementation for the robot operating system," in *Intelligent Autonomous Systems 13* (E. Menegatti, N. Michael, K. Berns, and H. Yamaguchi, eds.), pp. 335–348. Cham: Springer International Publishing, 2016.

[45] K. Mohta, M. Watterson, Y. Mulgaonkar, S. Liu, C. Qu, A. Makineni, K. Saulnier, K. Sun, A. Zhu, J. Delmerico, et al., "Fast, autonomous flight in gps-denied and cluttered environments," *Journal of Field Robotics*, vol. 35, no. 1, pp. 101–120, 2018.

[46] X. Kan, H. Teng, and K. Karydis, "Online exploration and coverage planning in unknown obstacle-cluttered environments," *IEEE Robotics and Automation Letters*, vol. 5, no. 4, pp. 5969–5976, 2020.

[47] X. Kan, T. C. Thayer, S. Carpin, and K. Karydis, "Task planning on stochastic aisle graphs for precision agriculture," *IEEE Robotics and Automation Letters*, vol. 6, no. 2, pp. 3287–3294, 2021.

[48] M. Campbell, A. Dechemi, and K. Karydis, "An integrated actuation-perception framework for robotic leaf retrieval: Detection, localization, and cutting," in *2022 IEEE/RSJ International Conference on Intelligent Robots and Systems (IROS)*, Kyoto, Japan, pp. 9210–9216, 2022. https://doi.org/10.1109/IROS47612.2022.9981118

[49] K. Ye, G. J. Correa, T. Guda, H. Teng, A. Ray, and K. Karydis, "Development and testing of a novel automated insect capture module for sample collection and transfer," in *2020 IEEE 16th International Conference on Automation Science and Engineering (CASE)*, Hong Kong, China, pp. 1063–1069, 2020. https://doi.org/10.1109/CASE48305.2020.9216816

[50] N. Koenig and A. Howard, "Design and use paradigms for gazebo, an open-source multi-robot simulator," *IEEE/RSJ International Conference on Intelligent Robots and Systems (IROS)(Cat. No. 04CH37566)*, vol. 3, pp. 2149–2154, 2004.

3 Electrical Tractors for Autonomous Farming

Redmond R. Shamshiri

3.1 INTRODUCTION

Electric tractors have gained popularity in recent years due to their numerous benefits compared to traditional diesel tractors. Efficient battery technology is becoming increasingly available and suitable for use in compact tractors on farms of the future. One of the main reasons for the rise of e-tractors is the growing concern for the environment and the need to reduce greenhouse gas emissions. Increasingly stringent demands for reducing noise and exhaust emissions are also driving the development of quiet and powerful e-tractors that are ideal for smaller farms and utility operations. These machines reduce carbon emissions and can perform many tasks of diesel tractors without generating noise or exhaust. They are quiet and powerful for their size, operating very similarly to diesel-powered tractors in terms of controls, hydraulics, and three-point assembly.

The first electrical tractor was introduced in the 1970s and was powered by lead-acid batteries [1, 2]. These prototypes were primarily used for light-duty work such as mowing and spraying. In the 1990s, advancements in battery technology and electric motor design led to the development of more powerful electric tractors that were suitable for heavy-duty work such as plowing and tilling [3]. These tractors were equipped with larger batteries and more efficient electric motors, allowing them to operate for longer periods and with greater power. However, the high cost of batteries and the limited range of e-tractors were significant barriers to their widespread adoption in large agricultural fields. It was not until the early 2000s that the technology had advanced enough to make e-tractors a viable alternative to diesel tractors. Several companies manufacture e-tractors for agricultural use, with some models capable of operating for up to 8 hours on a single charge. Additionally, models like Solectrac (Solectrac, Windsor, CA) can be charged using solar panels, further reducing operating costs and environmental impact.

The objective of this chapter is to provide an overview of e-tractors in agriculture, including their history, development, current usage, and potential for transforming the agricultural industry. The chapter will explore the advantages and disadvantages of e-tractors, their impact on productivity and sustainability in agriculture, as well as the challenges and opportunities for farmers interested in adopting this technology.

DOI: 10.1201/9781003306283-3

3.2   BACKGROUND AND APPLICATIONS

Electrically powered agricultural machinery has emerged as a transformative force in the agricultural industry, representing a shift towards sustainable and environmentally conscious farming practices. E-tractors are projected to be a major driver in the agricultural industry's adoption of contemporary farming methods. The global electric farm tractor market is anticipated to experience growth driven by the growing need for agricultural digitalization and automation, coupled with an increasing demand for eco-friendly solutions aimed at reducing emissions and mitigating pollution. According to a report from a business consulting company called BIS Research, the global electric farm tractor market was valued at $116.5 million in 2020, and it is expected to grow at a compound annual growth rate (CAGR) of 11.1% to reach $218.9 million by 2026.

A typical e-tractor has a battery runtime between 3–10 hours depending on the load, which can be charged in under 10 hours using the standard 220/110 AC connection outlets [4]. They can accept all category I and II implements on their hydraulically controlled rear three-point hitch and come with 540 RPM power take-off shaft. The price of e-tractors can vary widely depending on the specific model, features, and manufacturer, but generally is in the range of 20,000 to over 250,000 USD [5]. The average price for a 25 HP category e-tractor ranges between 20,000 and 60,000 USD. The average weight of e-tractors can vary depending on the model, size, and power of the tractor. Generally, e-tractors tend to be lighter than traditional diesel tractors due to the absence of a heavy engine and transmission. However, the weight of an e-tractor can still vary widely depending on its intended use and features. For example, smaller e-tractors designed for use in small fields or orchards may weigh around 500 kg to 1000 kg, whereas larger models used for commercial farming can weigh 5000 kg or more.

Similar to conventional diesel engine tractors, farmers can use e-tractors for a wide range of agricultural operations, but they are particularly well-suited for tasks that require high torque at low speeds, such as tilling, cultivating, and plowing. This is because electric motors can provide high torque output at low speeds, making them more practical for heavy workloads. Additionally, e-tractors can be used for hauling, mowing, and spraying, making them ideal for use in small gardens, orchards, or vineyards, where their smaller size and quiet operation can be beneficial. If farmers already have solar panels on their farms, they will experience close to zero cost for battery charges. Even without solar panels, their fuel costs should be reduced depending on local electricity costs. E-tractors are being successfully employed in various ways across different types of farms, including vineyards using Monarch Tractors, orchards using Weidemann 1190e for mowing, dairy farming using the New Holland T6.180 to move feed and clean barns, livestock production using Kioti MECHRON 2240, and small or medium arable lands using Fendt e100 Vario for planting, harvesting, and transporting crops. Figure 3.1 shows the Weidemann 1190e articulated steering electrical tractor equipped with a custom-built electrical mower that is used in the SunBot project for manual and autonomous mowing [6]. These e-tractors can operate in the field for approximately 5–10 hours, depending on the type of use and speed. They can be equipped with various attachments, such as electrical mowers or loaders,

FIGURE 3.1 An articulated steering electrical tractor manufactured by Weidemann equipped with a custom-built electrical mower used in the SunBot project.

making them suitable for a range of farming applications without the need for engine oil changes. These applications include mowing, soil preparation for planting bushes, seeding, and even pulling trailers or moving materials. Due to their compact size, these machines function very effectively in small spaces where larger tractors may struggle to work, and they can operate without generating noise and exhaust pollution near barns or farmhouses. Farmers are encouraged to use these machines and evaluate their performance, which will enable manufacturers to scale up production and reduce prices. Smaller e-tractors designed for use in small fields or gardens typically have a power output of around 15–20 horsepower (11–15 kilowatts), whereas larger models intended for commercial farming can have a power output of 100 horsepower (75 kilowatts) or more.

Research efforts in the development of driverless e-tractors capable of autonomously navigating within unstructured orchards and performing deterministic tasks have significantly increased over the past two decades, enhancing autonomy and economic feasibility for an energy transition [7]. Examples include design of a heavy-duty platform for autonomous navigation in kiwifruit orchards [8], vision-based control for self-steering tractors [9], design of an autonomous charging station for e-tractors [10], and research and experiments on automatic navigation control technology of intelligent e-tractors [11]. Some of the e-tractors, such as John Deere GridCON [12], have been equipped with built-in cameras and artificial intelligence and are available to work in row-crops semi-autonomously using joystick control and touchscreen display. Most of the modern e-tractors come with mobile apps [13] that allow farmers to compare diesel fuel and electricity costs over time, calculate emissions reduction, estimate the cost saving of using multiple e-tractors, and compare the maintenance fees. Such information can be shared within the community or used for further data analysis.

3.3 BENEFITS OF USING E-TRACTORS

E-tractors offer safer and cleaner field operations and make a significant contribution to zero-emission regenerative agriculture. Unlike diesel tractors, which emit harmful pollutants such as CO_2, NO, and particulate matter, e-tractors have a lower carbon footprint, produce zero emissions, and do not generate toxic gases or cause air pollution when operated [14], hence contributing to sustainable agriculture practices that prioritize climate change and conservation. This can have a positive impact on soil health, water quality, and overall ecosystem health. By using renewable energy sources such as wind or solar power to charge the batteries, the carbon footprint of e-tractors can be reduced even further. From a technical perspective, e-tractors have more consistent power delivery, better control and precision [15], improved operator comfort, reduced operating and maintenance cost, and longer lifespan [13]. Maintaining consistent power delivery means they can maintain their performance even under heavy loads [16]. In addition, with e-tractors, having instant torque is possible [17], enabling them to start operating at maximum power immediately. These options result in faster acceleration, better performance, increased productivity, and faster completion of tasks. The ability for more precise speed control offer better maneuverability, making e-tractors ideal for precision farming techniques such as GPS-guided navigation. This is required and useful for some tasks such as precise planting, fertilization, and harvesting, which can lead to increased yields and better quality crops.

The operating and maintenance costs of e-tractors are significantly lower, something that is proving true in electric cars, because the engine has fewer moving parts compared to dozens in a diesel tractor, less wear and tear on the engine, and less exposure to emissions that can cause damage to the machines' components [18]. As a result, they have longer lifespans, which can help farmers reduce the long-term costs of purchasing new equipment. Additionally, they produce less heat and vibration, making them quieter than diesel tractors. This provides farmers with more comfortable driving experiences, allowing them to use the machines for extended periods. The reduction in noise pollution can also benefit nearby residents and wildlife, as loud noises can disrupt their habitats and behavior. Apart from being cleaner and quieter, e-tractors are more energy-efficient compared to their diesel counterparts. This translates to lower operating costs and reduced reliance on fossil fuels. With the option for solar-charged battery swaps, farmers can change the battery in the field in under 10 minutes, resulting in significant cost and time savings, especially during long tractor operation sessions. Figure 3.2 shows two models of e-tractors from two Indian manufacturers for emission-free farming.

FIGURE 3.2 Electrical tractors contributing to Indians' remote fields for emission-free farming, left: Cellestial E-mobility from Hyderabad startup (source: evreporter.com), and right: AutoNxt's autonomous tractor from a Mumbai Startup (source: inc42.com).

3.4   CHALLENGES AND LIMITATIONS WITH E-TRACTORS

The technologies used in e-tractors, especially in the battery management system, are on the cutting edge now, but they will likely become industry standard in the next one or two decades. The main concern with e-tractors is the range and reliability of their battery, which most manufacturers claim a battery life of 10 years pending 2500–3000 operation cycle and depth discharge of 80% [4]. The battery pack of the John Deere SESAM e-tractor is shown in Figure 3.3, which includes 182 cells of 3.7 v Lithium-Ionen battery cells in series connection, resulting in a maximum constant power of 133 kW. Most of the available e-tractors have a range of 4–5 hours, which may not be enough for large farms [18]. Farmers may need to invest in additional batteries or charging infrastructure to ensure their tractors have enough power to complete a full day's work. To overcome this challenge, research and development in battery technology is needed to improve the energy density, reduce the weight, and increase the lifespan of batteries. One of the solutions being developed is the use of swappable batteries that can be quickly and easily exchanged for fully charged ones [19]. This would reduce the downtime for charging and allow for continuous use of the e-tractor. Ghobadpour et al. [20] reported on a study with the goal of creating an Energy Management System (EMS) designed for a plug-in hybrid electric tractor (PHET) with the primary objectives of reducing fuel consumption and extending the operational range of the tractor. Their study covered the development of an extended-range, solar-assisted plug-in hybrid electric tractor tailored for light agricultural tasks.

The lack of charging infrastructure in rural areas can also be a major barrier to the adoption of e-tractors. Farmers may need to invest in charging infrastructure on their own or rely on public charging stations, which may be limited in availability. Some farmers are benefiting from existing solar panels on their farms to charge the batteries, leveraging renewable energy sources and reducing the need for grid electricity. The heavy weight of the batteries can affect the tractor's overall weight distribution and stability. To overcome this, e-tractors can be designed in a modular fashion, allowing farmers to customize their tractors to their specific needs. For example, a small tractor can be expanded with additional batteries and motors to create a larger tractor as needed. For the safety of the battery, commercial e-tractors must include features such as shutdown switches (located in easily accessible positions) on both sides of the tractor, direct wire communication with the battery system

FIGURE 3.3   Lithium-Ionen battery pack of John Deere SESAM, including 182 cells of 3.7 v battery cells in series connection, resulting in a maximum constant power of 133 kW (source: www.clubofbologna.org).

to shut down the power supply of the tractor when requested by the safety controller, easy access (via a connector) to ignition signals (similar to Clamp15 on the generator of combustion engines), and insulation monitoring equipment for the tractor's power circuit and auxiliary components.

One of the main limitations to the adaptation of e-tractors is their cost. They can be more expensive than their diesel counterparts due to limited supply, market availability, and the high cost of batteries. This can make the initial investment a barrier for many farmers to afford. Additionally, e-tractors may not be able to match the power output of diesel tractors, especially for heavy-duty tasks such as plowing and tilling. They also require different maintenance procedures compared to diesel tractors, and farmers may need to invest in training or new tools to properly maintain their equipment. In fact, addressing the challenges faced by e-tractors in agriculture will require a combination of technological innovation, infrastructure investment, government support, and manufacturers providing education and training for farmers. Advances in battery technology such as fast charging [21], high-energy density batteries [22], and solid-state batteries [23] can drive the development of more efficient, cost-effective, and environmentally friendly e-tractors.

3.5 AVAILABILITY OF E-TRACTORS

E-tractors can be classified based on their power units, which includes batteries, a combination of batteries and diesel engine (hybrid), or hydrogen cells. Battery-powered e-tractors are the most common type that are typically running on high-capacity lithium-ion batteries and can be charged using standard AC outlets, specialized charging stations, or solar panels. Some of the models that are available in the market include Kramer 5055e, Edison HTZ 3512, Solectrac e25, Fendt e100-vario, Weidemann 1160, Farmtrac FT25G, John Deere SESAM, and Rigitrac SKE 50, as shown in Figure 3.4. The main specifications of these battery-powered e-tractors were collected from their manufacturers' websites and are presented in Table 3.1. An example of a hybrid e-tractor is the New

FIGURE 3.4 Examples of the electrical tractors available in the market: (a) Kramer 5055e, (b) Edison HTZ 3512, (c) Solectrac e25, (d) Fendt e100-vario, (e) Weidemann 1160, (f) Farmtrac FT25G, (g) John Deere SESAM, (h) Rigitrac SKE 50.

TABLE 3.1

Main Specifications of Some of the Available E-Tractors in the Market

| Manufacturer | Model | Power | | Maximum Speed | Battery Capacity | | | Running Time |
		(hp)	(kW)	(km/h)	V	Ah	kWh	(h)
KRAMER	5055e	40	35	17	80	416	33	3
Edison	HTZ 3512	35	26	40	48	500	24	NA
Solectrac	e25	25	18.4	40	72	300	21.6	3 to 6
Fendt	e100-vario	67	50	50	650	153	100	5
Weidemann	1160	25	18.4	20	48	240	11.5	1 to 2
FARMTRAC	HST 25	25	18.4	30	72	300	21.6	NA
John Deere	SESAM	174	130	30	400	250	100	3
Rigitrac	SKE 50	67	50	40	400	250	100	3

Holland T6, which can switch to a diesel engine when needed [24]. Hydrogen fuel cell e-tractors [25] are powered by hydrogen fuel cells, which generate electricity by combining hydrogen and oxygen. These tractors produce no emissions other than water vapor, and they have a longer range than battery electric tractors; however, they are more expensive and require specialized infrastructure for refueling.

The Kramer 5055e is produced by Kramer-Werke GmbH, a German company that has been manufacturing agricultural machinery since 1927. It is a highly efficient and versatile electric tractor that is well-suited for use in agriculture and other industries. One of the main features of the Kramer 5055e is its electric drive system. It is powered by a 60-kilowatt electric motor that is connected to a high-capacity lithium-ion battery pack. The battery pack has a capacity of 86 kilowatt-hours and can be charged using a standard 400-volt industrial outlet. The tractor has a maximum speed of 30 km/h and can operate for up to 5 hours on a single charge, depending on the application. It can also be equipped with a range of features that are designed to improve its performance and efficiency, such as intelligent EMS that optimizes the use of the battery pack to ensure maximum range and performance. This e-tractor benefits from regenerative braking (RB), which helps recharge the battery pack when the machine is slowing down or stopping. In addition, the Kramer 5055e has a range of agricultural attachments and implements that are designed to improve its versatility and usefulness on the farm. These include a front loader, rear hitch, and a range of cultivators, plows, and other implements.

The Edison HTZ 3512 is produced by Edison Motors, a company based in Croatia that specializes in electric vehicles. It is powered by a 35-kilowatt electric motor that is connected to a high-capacity lithium-ion battery pack with a capacity of 90 kilowatt-hours and can be charged using a standard 400-volt industrial outlet. It has a maximum speed of 40 km/h and can operate for up to 8 hours on a single charge, depending on the application. The Edison HTZ 3512 is also equipped with an intelligent energy management system (IEMS) and regenerative braking (RB), with a range of agricultural attachments, including a front loader, rear hitch, and a range of cultivators, plows, and other implements.

Solectrac is produced by a California-based company for small farms and homesteads. The e25 is equipped with a 25 hp electric motor and can reach a maximum speed of 40 km/h. It has a 60 kWh lithium-ion battery and a payload capacity of up to 2500 pounds, and can tow up to 4000 pounds. The battery capacity of the Solectrac e25 can vary depending on the specific configuration and options chosen by the customer, which is designed to provide up to 4 hours of continuous operation on a single charge using a standard 240-volt outlet or a Level 2 charging station. The tractor has a compact size and a tight turning radius, making it suitable for use in small or tight spaces. It also has a three-point hitch system and a rear PTO for attaching various agricultural implements, such as mowers, cultivators, and seeders. It has been tested on several farms in the US and has received positive feedback from farmers. The Solectrac e25 can be attached to a range of agricultural attachments, such as a front loader, rear hitch, and a range of cultivators, plows, and other implements.

In 2018, German tractor manufacturer Fendt introduced the e100 Vario, an e-tractor designed for light to medium-duty work with a maximum speed'of 50 Km/h which can vary depending on the specific model and operating conditions. It is the company's first all-electric tractor and is designed to provide farmers with a cost-effective and environmentally friendly alternative to traditional diesel-powered tractors. It has a 50 kW electric motor that is powered by a 100 kWh high-capacity lithium-ion battery pack and a range of up to 5 hours and can operate at full power for up to 2 hours. The Fendt e100 Vario can be charged using a standard 400-volt plug or a fast charging station. It has customizable features such as adjustable driving modes and a touchscreen display, which can help increase operator comfort and productivity. The Fendt e100-vario tractor can also be charged using renewable energy sources, such as solar panels, making it even more environmentally friendly.

The Weidemann 1160 is an articulated steering eHoftrac produced by Weidemann, a German manufacturer of agricultural and construction equipment. The 1160 eHoftrac is part of Weidemann's line of Hoftrac tractors, which includes several other models that are powered by diesel engines. It is powered by a 48-volt electric motor with a maximum output of 14.8 kW (approximately 19.8 horsepower) that is connected to a 12 kWh high-capacity lithium-ion battery pack and can be charged using a standard 230-volt household outlet. The tractor has a maximum speed of 20 km/h and can operate for up to 6 hours on a single charge, depending on the application. One of the main features of the Weidemann 1160 eHoftrac is its compact size and maneuverability due to its articulated steering. It is designed for use in small farms and other applications where space is limited. The Weidemann 1160 eHoftrac is equipped with a range of attachments, including a front loader and fork, and a range of safety features, such as an anti-rollover system and a safety cab that protects the operator from falling objects.

Farmtrac FT25G is a compact and versatile e-tractor with a strong and durable chassis that is designed to withstand the toughest farming conditions. It is supported by a 15 kW induction motor and a 72 V, 300 Ah lithium-ion battery, which can be charged through any standard AC charging outlets in under 5 hours. The manufacturer of the Farmtrac 25G claims that this e-tractor will reduce the farmer's operating costs by more than 50%. Its compact size makes it easy to maneuver in tight spaces, while its high ground clearance ensures that it can handle rough terrain with ease. It features a hydrostatic transmission system that allows for smooth and easy shifting between gears. This system ensures that the tractor maintains a constant speed and power output, making it ideal for a wide range of farming applications. The Farmtrac FT25G is compatible with various implements, including a front blade, rotavator, and wood chipper.

In 2016, John Deere introduced a high-horsepower e-tractor prototype called the Sustainable Energy Supply for Agricultural Machinery (SESAM), a 130 kW tractor powered by a 100 kWh lithium-ion battery pack (shown in Figure 3.5) that can be charged in 3 hours and has a working time range of 4 hours. The SESAM tractor is part of John Deere's ongoing efforts to reduce the environmental impact of large-scale farming and increase sustainability. It weighs about the same as a traditional diesel-powered tractor of similar size with the same power and efficiency, and can be fitted with a range of implements such as plows, cultivators, and seed drills. The Deere eAutoPowr transmission employs a unique approach, utilizing two high-power electric machines instead of the conventional hydrostatic unit to achieve the necessary gear ratio. This innovative configuration enables the generation of an additional 100 kW of electric power for external use, thereby enhancing overall efficiency and performance. In the Deere eAutoPowr transmission, one electric machine is connected to the engine, while the other is linked to the summing planetaries, allowing for infinitely adjustable wheel speeds. This setup offers traction assistance for the slurry implement, reducing field compaction and minimizing soil damage through reduced slip. This gives farmers the flexibility to operate in less-than-ideal conditions, ultimately increasing productivity and expanding the available working hours within tight timeframes.

The Rigitrac SKE 50 is a powerful and versatile electric tractor built by the Swiss manufacturer Hürlimann that has demonstrated impressive performance, reliability, and eco-friendliness for use in sensitive environments such as orchards, vineyards, and urban areas. It weighs 4000 kg, has a power output of 50 kW, maximum torque of 400 Nm, and top speed of 50 km/h. The motor is paired with a high-capacity 100 kWh lithium-ion battery that provides enough energy for up to 6 hours of continuous operation, or 120 km range on a single charge. The SKE 50 also features a range of advanced features such as fully adjustable suspension system that provides excellent stability and maneuverability on uneven terrain, as well as an intuitive touchscreen interface that allows for easy monitoring and control of all tractor functions.

Other e-tractors or electrical vehicles that can be used for farming include New Holland T4 and T6, Monarch e-tractor, Weedingtech, Kubota X tractor, Case IH Autonomous Concept Vehicle, and Kioti Mechron 2240 Electric. The New Holland T4 utility electric self-driving e-tractor is a four-wheel drive prototype model with maximum speed of 40 km/h, 120 hp, and maximum torque of 440 N.m, developed in collaboration with Monarch tractor and CNH industrial. The New Holland T6 is a hybrid e-tractor that combines an electric motor with a diesel engine. It is designed to operate primarily on electric power, but can switch to the diesel engine if needed. The T6 has a range of up to 3 hours on electric power. Monarch Tractor is a startup company that has developed a fully electric tractor designed for small to mid-sized farms. The tractor has a 70 kWh battery pack and a range of up to 10 hours. It can be charged using a standard 110-volt plug or using solar panels. The Monarch Tractor also features autonomous driving technology, which can be used for precision agriculture. Weedingtech is a UK-based company that has developed an electric weed-killing machine called the Foamstream. The Foamstream is mounted on an e-tractor and uses hot foam to kill weeds. The Foamstream system is 100% chemical-free and is powered by renewable energy. The Kubota X tractor (shown in Figure 3.5) is a concept tractor that is designed to be autonomous and electric. It is powered by four electric motors and can operate for up to 10 hours on a single charge. The Case IH Autonomous Concept Vehicle (shown in Figure 3.5) is an autonomous electric tractor that is designed to operate without human intervention. It is powered by a battery and has a range of up to 14 hours. The Kioti Mechron 2240 Electric is a utility vehicle that runs on a 48-volt system and features a 4 kW AC motor. It is designed for a variety of applications, including agriculture and landscaping, and has been demonstrated to be reliable and durable, capable of handling tough terrain and heavy loads with a tow capacity of up to 580 kg and a payload capacity of up to 650 kg. It has a maximum speed of 40 km/h and a range of up to 50 miles on a single charge, depending on the terrain and conditions. It features a 6-inch ground clearance and has a cargo capacity of up to 500 kg.

3.6   E-TRACTORS FOR HIGH-DENSITY ORCHARDS

One of the main requirements for the operation of e-tractors inside unstructured and semi-structured orchards with high-density plants, such as the one shown in Figure 3.6, is to navigate through narrow row-spaces that are between 1.5–2.5 m while avoiding collision with bushes on the sides. For the case of berry orchards, a functional

FIGURE 3.5 Examples of e-tractor concepts for the farms of the future: (a) Case IH Autonomous Concept Vehicle (image source: caseih.com), (b) a conceptual e-tractor from Class, (c) an autonomous e-tractor with fuel cell drive developed by the Chinese National Institute of Agromachinery Innovation and Creation (image source: auto-motor-und-sport. de), (d,e) two conceptual designs from Valtra group (image source: valtra.com), and (f) Kubota X tractor (image source: kubota.com).

e-tractor should be able to navigate autonomously between these rows with an accuracy of 5–10 cm from the plants while maintaining an ideal speed of 5–8 km/h. This is, however, an extremely challenging task due to the structural dynamics of the growing orchard, including land topology and the variation in the planting systems.

View of a berry orchard with variation in row spacing is shown in Figure 3.7, in which a possible solution is a universal two-wheel-drive e-tractor that operates on a $LiPO_4$ battery with all-electric drive, has a power range of 20–30 kW, overall width of 1.3 –1.5 m, and a total weight that does not exceed 2500 kg. Such an e-tractor should have front and rear three-point linkage C1 and an AEF electric power connector to power an electrically driven implement such as a double-blade mower (400 kg/10 kW). CANBUS communication is preferred for autonomous driving control signals (i.e., steering and speed), and ISO-BUS protocol (ISO-11783 standard) for data exchange between the tractor and mounted implements. Some of the messages that are exchanged include control messages for autopilot, command signals, feedback values, status messages, ground speed message, brake control signal, battery status signal, power consumption status of the tractor, electric power take-off control, position control of the linkage, speed control of the PTO, distance between the tractor and obstacles, human detection, and alarms. Table 3.2 provides a summary of the main requirements for an e-tractor that can perform autonomous mowing in such orchards.

Field experiments with GPS navigation inside the orchard shown in Figure 3.6 and Figure 3.7 revealed that absolute measurements were sometimes insufficient to

FIGURE 3.6 View of a high-density orchard with challenging narrow row spaces for navigation of e-tractors.

FIGURE 3.7 Berry orchard with variation in row spacing between bushes.

TABLE 3.2
Main Requirement for an Electrical Tractor to Operate in a Berry Orchard

Specification	Requirement
Tractor type and wheel drive	Universal, two-wheel drive (4WD optional)
Power	20–30 [kW] or 25–40 [hp]
Tractor overall width (without cabin)	Maximum 1.37 [m]
Track width of rear wheels	Between 1.1–1.35 [m]
Option for track width of rear wheels	Adjustable to tire sets 1.3, 1.4, 1.45 [m]
Track width of front wheels	Between 1.2–1.35 [m], minimum 1.2 [m]
Wheelbase	Maximum 2.1 [m]
Ground clearance	Minimum 0.27 [m]
Optional for front and rear wheel	Wheel positions adjustable on axle
Tractor steering type	Front axle, Ackerman steering
Interface for steering command	Open interface: autopilot to tractor controls (CAN BUS)
Steering angle	Standard 55°
Tractor turning radius with brake	3.5 [m]
Tractor turning radius without brake	4.0 [m]
Front axle articulation	$\approx 21°$
Tractor operating weight (including battery)	Maximum 2500 [kg]
Total weight (without battery)	Maximum 1600 [kg]
Permitted towed weight implement	Maximum 3000 [kg]
Maximum forward speed	22–28 [km/h]
Maximum reverse speed	22–28 [km/h]
Rear three-point hitch type	Standardized hydraulic three-point hitch
Linkage dimension, front and rear hitch	Category 2 three-point linkage
Payload rear hitch	Minimum 600 [kg]
Payload front hitch	Minimum 400 [kg]
Embedded functionality of three-point hitch	Float position required (optional auto draft and depth control)
Mechanical power take-off (PTO), rear	Requirement
PTO type, control	TYPE 1, Independent
PTO speed/power	540 [RPM]/23.60 [hp]
PTO shaft dimension/teeth	35 [mm]/6 teeth
Front PTO	Optional
Working pressure of the hydraulic power take-off	16 [MPa]
Flow rate of the hydraulic system of the attachment	20 [l/min]
Number of interfaces for electrical power take-off	2 electric power interfaces for implement power supply

TABLE 3.2 *(Continued)*
Main Requirement for an Electrical Tractor to Operate in a Berry Orchard

Specification	Requirement
Position of the electrical power take-off interfaces	1 x front and 1 x rear, near implement attachment
Electric power interface	AEF-compliant electric power interface
System voltage (battery)	DC, 400 [V], 40 [A] or 48 [V], 40 [A]
Connector type	Socket according to AEF standard (or similar) containing:
Connector fault current detection	insulation monitor installed in socket or on tractor
Connector shielding	Shielded housing
Connector interlock	Electric interlock system
Tear-proof	Mechanical interlock or tear-proof connector system
Accessories supply: number of interfaces	2 x low voltage supply interfaces for accessories
Position	Cabin
Power range of electrical socket	12 [V]/60 [A] and 24 [V]/30 [A]
Implement data communication	CANBUS
Number and position of the communication port	1 front and 1 rear, near implement attachment
Communication port	ISOBUS—ISO 11783 plugs
Low-volt electrical power supply	12 [V], 60 [A]—ISO 11783 plugs
Data communication protocol with implement	ISO 11783 (ISOBUS) or compatible BUS system
Battery type	Li-Lon, Li-Po, or LiFePO$_4$ (3.2V 30 Ah LiFePO$_4$) or better
Battery recharge capacity	After 2500 cycles, at least 80% of the initial capacity
Charging current of the battery	AC 52 [A] at 380 [V], AC 91 [A] at 220 [V], and DC 65 [A]
Charging time of the battery	Less than 4 hours
Battery working time when loading at least 90%,	Minimum 4 hours
Supply voltage of the charging station	AC 380 [V]/220 [V], DC 400 [V]

accurately determine the heading and position corrections [26], which causes the GPS-based navigation to be interrupted due to poor signal reception and faulty or noisy readings. Therefore, it is essential to incorporate different sensing solutions such as depth cameras, multi-channel distance detection sensors, and Light Detection and Ranging (LiDAR) devices to provide e-tractors with autonomy. The best mounting positions for these sensors can be determined using simulated models, such as the one shown in Figure 3.8. Simulation models also make possible analyzing kinematic behavior of the vehicle at the row-end turnings according to the actual dimensions of the farmer's fields. The sensors mentioned in Figure 3.8 are required to withstand harsh field conditions, have a flexible control design with interchangeable and compatible components, and benefit from a software program that can be adjusted by farmers for different field applications.

Data fusion and multiple perception solutions are usually employed to assist the existing GPS-based navigation and to improve the reliability of the operation. The communication between multiple sensors and controllers are realized via two separate CANBUS lines that are bridged together via a ROS-based vehicle control unit, as shown in Figure 3.9. It should be noted that the development of an effective autonomous navigation system for e-tractors that can have reactive behavior and reflexive responses to react to an unknown situation and to make rapid decisions such as changing speed or steering angle requires extensive experiments with different scenarios. To accelerate this pace, simulation of the components in virtual environments can provide an affordable and reliable framework for experimenting with different

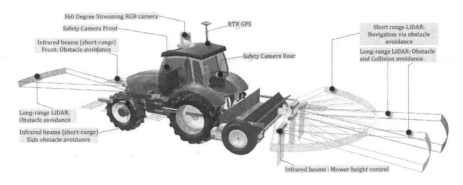

FIGURE 3.8   A simulated e-tractor with various sensors for autonomous mowing in berry orchards.

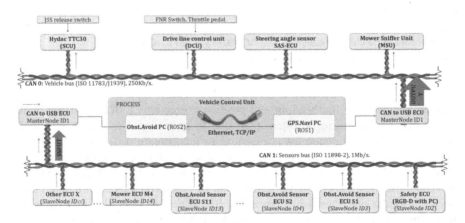

FIGURE 3.9   Schematic description of the communication between multiple sensors and controllers for an autonomous e-tractor with electrical mower and collision avoidance capabilities.

sensing and acting mechanisms in order to verify the performance of the robot to interact with highly variable dynamic scenarios.

3.7   ACCELERATING THE ADOPTION OF E-TRACTORS

Governments worldwide are actively advocating for the implementation of advanced agricultural machinery within the industry. Promoting the adoption of electric farm tractors through increased incentives for farmers represents a promising strategy to stimulate their use in the present context. This initiative is poised to foster the utilization of electric farm tractors, thereby enhancing agricultural productivity. Before investing in an e-tractor, farmers are expected to assess their farm's specific requirements, taking into consideration factors such as land size, terrain, and types of crops to maximize machine efficiency and the field machine index. In terms of financial planning, farmers should examine available resources, including government incentives, subsidies, and grants designed to encourage sustainable farming practices. Such initiatives can significantly reduce the initial costs associated with adopting e-tractors. Moreover, investing in charging infrastructure is necessary to ensure uninterrupted operations, which may involve establishing on-site charging stations or collaborating with nearby charging facilities. Some models of e-tractors require specific charging stations, and farmers may need to upgrade their existing electrical infrastructure or consider installing solar panels or wind turbines to generate renewable energy to power their tractors. Farmers with the opportunity to set up rooftop solar installations on structures such as barns, sheds, and various other buildings can have a sustainable, cost-effective source of electricity for charging their e-tractors as well as supporting energy-intensive operations such as processing, packaging, and refrigeration. Certain regions also offer net-metering initiatives that enable farmers to sell surplus energy to the grid, thereby increasing their income. Additionally, ground-mounted solar systems present another viable choice for both on-grid and off-grid power solutions situated in close proximity to the power-consuming loads. Figure 3.10 shows a charging station from Paired Power with solar panels installed on a canopy for charging e-tractors.

FIGURE 3.10   A charging station from Paired Power for e-tractors featuring canopy with solar panels (source: Paired Power).

To ensure the long-term optimal performance of e-tractors, farmers should gain proficiency in e-tractor operation and maintenance, establish a well-structured service schedule, and plan for necessary repairs. Routine servicing not only extends the machine's lifespan but also ensures consistent functionality and reduces downtime. Policymakers can support the adoption of e-tractors by investing in the development of rural charging infrastructure to ensure easy access to charging stations, as well as by expanding incentive programs, such as tax incentives, subsidies, and grants, to make eco-friendly farming machinery more accessible to farmers across various scales of operation. Additionally, allocating funding and resources to promote research and development in e-tractor and battery technology can contribute to efficiency enhancements and cost reduction. Policymakers can further promote this technology and accelerate the transition toward a greener, more productive, and environmentally responsible agricultural industry by supporting training and educational programs and initiating collaborations among governmental bodies, agricultural organizations, and e-tractor manufacturers.

3.8 CONCLUSION

Electric tractors mark the start of a fresh chapter in agricultural technology. Farmers are perpetually in search of approaches that can streamline and economize their cultivation practices, and technology presents a viable solution. E-tractors hold a distinct advantage in the digitalization of agriculture due to their capacity to reduce fuel consumption and their flexibility for implementing autonomy and more precise control. Powered by batteries, they simplify and economize agricultural operations. The emergence of e-tractors holds the potential to revolutionize the existing agricultural landscape, potentially amplifying productivity and efficiency by a large factor. With the advances in wireless communication, e-tractors can be connected to the internet and other farm equipment using long-range wireless transmitters and the Internet-of-Things (IoT) devices in order to provide farmers with teleoperation control, live monitoring of their tractors' locations, battery status, and remote management of field operations. This can result in generating valuable data on crop growth, soil health, and weather conditions, which are used by decision support systems to make knowledge-based decisions and optimize farming practices. Successful development of such a system requires a proof-of-concept via extensive validation tests that can be accelerated with the digital representation of the sensors, dynamic models of the e-tractor, and virtual replicas of the orchard.

While the upfront cost of an e-tractor may be higher than that of a diesel model, the lower cost of electricity compared to diesel fuel, as well as the lower maintenance costs due to fewer moving parts, can make them a more cost-effective choice in the long run. In addition, advancements in battery technology are making e-tractors more practical for use in agriculture. Battery technology is improving rapidly, and newer batteries can provide longer operating times and faster charging times than older models did. This means that electric tractors are becoming more viable for use in large-scale agricultural operations. The surveyed literature shows that the future of electric tractors in agriculture is very bright. As more farmers become aware of the benefits of eco-friendly farming equipment, and as technology continues to improve, it is expected to see a growing number of e-tractors on farms around the world. If successfully integrated and implemented, fully autonomous e-tractors can play a key role in reducing agricultural production costs by decreasing the number of human workforces that are currently engaged in performing repetitive tasks.

REFERENCES

[1] A. Malik and S. Kohli, "Electric tractors: Survey of challenges and opportunities in India," *Materials Today: Proceedings*, vol. 28, pp. 2318–2324, 2020.

[2] Y. Ueka, J. Yamashita, K. Sato, and Y. Doi, "Study on the development of the electric tractor: Specifications and traveling and tilling performance of a prototype electric tractor," *Engineering in Agriculture, Environment and Food*, vol. 6, no. 4, pp. 160–164, 2013.

[3] R. R. Melo, F. L. M. Antunes, S. Daher, H. H. Vogt, D. Albiero, and F. L. Tofoli, "Conception of an electric propulsion system for a 9 kW electric tractor suitable for family farming," *IET Electric Power Applications*, vol. 13, no. 12, pp. 1993–2004, 2019.

[4] D. L. Bessette, D. C. Brainard, A. K. Srivastava, W. Lee, and S. Geurkink, "Battery electric tractors: Small-scale organic growers’ preferences, perceptions, and concerns," *Energies*, vol. 15, no. 22, 2022.

[5] T. Woopen, A. Gronewold, H. Adam, and S. Hammes, "Marketable electric powertrain concepts for smaller tractors," *ATZoffhighway Worldwide*, vol. 11, no. 2, pp. 8–13, 2018.

[6] C. Weltzien and R. R. Shamshiri, "SunBot: Autonomous nursing assistant for emission-free berry production, general concepts and framework," *LAND.TECHNIK AgEng*, pp. 463–470, 2019.

[7] H. H. Vogt et al., "Electric tractor system for family farming: Increased autonomy and economic feasibility for an energy transition," *Journal of Energy Storage*, vol. 40, p. 102744, 2021.

[8] M. H. Jones et al., "Design and testing of a heavy-duty platform for autonomous navigation in kiwifruit orchards," *Biosystems Engineering*, vol. 187, pp. 129–146, 2019.

[9] E. Vrochidou, D. Oustadakis, A. Kefalas, and G. A. Papakostas, "Computer vision in self-steering tractors," *Machines*, vol. 10, no. 2. 2022.

[10] E. H. Harik, "Design and implementation of an autonomous charging station for agricultural electrical vehicles," *Applied Sciences*, vol. 11, no. 13, 2021.

[11] X. Cai, W. Fan, Y. Wang, and Y. Qian, "Research and experiment on automatic navigation control technology of intelligent electric tractor," in *Proc. SPIE*, 2022, vol. 12349, p. 1234916.

[12] S. Karthik, "John Deere GridCON autonomous electric tractor," 2021 [Online]. Available: https://electricvehicles.in/john-deere-gridcon-autonomous-electric-tractor/. [Accessed: 12-Dec-2022].

[13] G. Steinberger, M. Rothmund, and H. Auernhammer, "Mobile farm equipment as a data source in an agricultural service architecture," *Computers and Electronics in Agriculture*, vol. 65, no. 2, pp. 238–246, 2009.

[14] A. Olaluwoye, "Harnessing electric tractors for sustainable farming in Ontario," York University, Toronto, Ontario, Canada, 2020. Available at: https://yorkspace.library.yorku.ca/items/c5c991ff-f34b-4108-a1d0-e8ddd2fe2923

[15] Y. An, L. Wang, X. Deng, H. Chen, Z. Lu, and T. Wang, "Research on differential steering dynamics control of four-wheel independent drive electric tractor," *Agriculture*, vol. 13, no. 9, 2023.

[16] C. R. Gade and R. S. W, "Control of permanent magnet synchronous motor using MPC–MTPA control for deployment in electric tractor," *Sustainability*, vol. 14, no. 19, 2022.

[17] G. C. S. Reddy, S. Deole, M. More, W. Razia Sultana, and A. Chitra, "Analysis of load torque characteristics for an electrical tractor," *Smart Grids and Green Energy Systems*, pp. 263–283, 2022.

[18] F. Mocera, A. Somà, S. Martelli, and V. Martini, "Trends and future perspective of electrification in agricultural tractor-implement applications," *Energies*, vol. 16, no. 18, 2023.

[19] Z. Wu, J. Wang, Y. Xing, S. Li, J. Yi, and C. Zhao, "Energy management of sowing unit for extended-range electric tractor based on improved CD-CS fuzzy rules," *Agriculture*, vol. 13, no. 7, 2023.

[20] A. Ghobadpour, H. Mousazadeh, S. Kelouwani, N. Zioui, M. Kandidayeni, and L. Boulon, "An intelligent energy management strategy for an off-road plug-in hybrid electric tractor based on farm operation recognition," *IET Electrical Systems in Transportation*, vol. 11, no. 4, pp. 333–347, 2021.

[21] Y. Liu, Y. Zhu, and Y. Cui, "Challenges and opportunities towards fast-charging battery materials," *Nature Energy*, vol. 4, no. 7, pp. 540–550, 2019.

[22 J. Li et al., "Toward low-cost, high-energy density, and high-power density lithium-ion batteries," *JOM*, vol. 69, no. 9, pp. 1484–1496, 2017.

[23] N. Boaretto et al., "Lithium solid-state batteries: State-of-the-art and challenges for materials, interfaces and processing," *Journal of Power Sources*, vol. 502, p. 229919, 2021.

[24] C. Valero Ubierna, P. Barreiro Elorza, M. Garrido Izard, and P. Diego, "Navegando a bordo de un New Holland T6 AutoCommand," *Vida Rural*, no. 378, pp. 14–26, 2014.

[25] H. Helms, M. Jamet, and C. Heidt, "Renewable fuel alternatives for mobile machinery," *Heidelberg: Institut für Energie-und Umweltforschung*, pp. 11–12, 2017.

[26] R. Shamshiri, C. Weltzien, I. Zytoon, and B. Sakal, "Evaluation of laser and infrared sensors with CANBUS communication for collision avoidance of a mobile robot," in *Proceedings International Conference on Agricultural Engineering. AgEng-LAND.TECHNIK 2022*, V. D. I. Wissensforum, Ed. Düsseldorf: VDI Verlag GmbH (0083-5560/978-3-18092406-9), 2022, pp. 121–130. https://www.vdi-nachrichten.com/shop/ageng-land-technik-2022/

4 Agricultural Robotics to Revolutionize Farming

Requirements and Challenges

Redmond R. Shamshiri, Eduardo Navas,
Jana Käthner, Nora Höfner, Karuna Koch,
Volker Dworak, Ibrahim Hameed,
Dimitrios S. Paraforos, Roemi
Fernández, and Cornelia Weltzien

4.1 INTRODUCTION

Modern agriculture is facing several challenges including climate change, market fluctuations, high labor costs, and declining participation of workforces that are willing to perform repetitive tasks under harsh field conditions. Facing these challenges necessitates improving the quality and efficiency of agricultural processes in order to remain competitive in the market. Farmers are seeking cultivation practices for producing more yields with higher quality at lower expenses in a sustainable way that is less dependent on the labor force. The innovative environment in agricultural robotics is characterized by the convergence of the advances in digital technology, including miniaturization of sensors [1], adoption of wireless communication [2], development of robust machine vision systems [3], and rapid expansion of artificial intelligence (AI) and machine learning algorithms [4]. These are widely considered to be promising solutions to revolutionize farming [5–9]. However, compared with industrial applications, the use of robots in agriculture has been slow to commerce mainly due to the high costs of research and development and lack of sufficient funding [10]. For most tasks, agricultural robotics are not yet robust and reliable enough to perform field tasks without human supervision [11]. This is due to the nature of the farming operations and the complex and variable requirements of both the environment and the tasks to be performed. For this reason, these robots are often operated by expert service providers who can make necessary adjustments in the fields. Along with the increasing number of startups that contribute to improving the technical robustness and trust-gaining ability of robots, governments worldwide have also initiated programs such as the Partnership for Robotics in Europe (SPARC) to support the development and adoption of agricultural robotics [10, 12]. In addition, governments are implementing other supportive measures to stimulate the development of agricultural robotics, including research grants, tax incentives, regulatory frameworks, and partnerships between academia, industry, and farmers, with the goal of creating an ecosystem that encourages innovation and reduces barriers to the adoption of agricultural robotics.

Robotic technologies can help address farming challenges as they can execute precise maneuvers, are highly flexible, and require less human labor, especially on repetitive and time-consuming tasks. Furthermore, the implementation of sustainable production systems depends largely on low production costs and high market prices to keep agricultural enterprises competitive. It is therefore important to evaluate the robotic systems to ensure that they meet the basic requirements for driving in perennial crops in terms of their advanced functions for applications in these cultures. The provision of suitable functionalities is considered as optional as a robot with a fitting chassis can be expanded by integrating additional sensor systems or tools. The requirements for field robots in crop care differ considerably from the requirements in stubble cultivation of the previous crop, fertilization, basic soil cultivation, seedbed preparation, and harvesting. Furthermore, it is generally advantageous to use one machine for more than just one working step. Therefore, a robotic system should be versatile and robust regarding its functionality.

In general, robots have several advantages over human labor, including higher accuracy and efficiency, better consistency and reliability, and lower operational costs, which are significant for farmers. Comprehensive research and development in agricultural robotics have been documented in a wide range of review papers [13–17] covering specific tasks such as phenotyping [18–20], arable farming [21], livestock farming [22], greenhouse horticulture [23], orchard management [24], forestry [25], and food processing [17]. Review papers also cover specific technologies used in agricultural robotics, such as computer vision [3, 26–28], active perception [29], path planning [30], and grasping and soft grasping [31, 32]. The majority of these studies have highlighted that for an agricultural robot to operate efficiently in harsh and unpredictable environments, and often under extreme weather conditions, it must be equipped with redundant sensing solutions to perceive its surroundings and be able to communicate and interact with other robots and machinery in the field.

The purpose of this study was to explore the recent advances and potential impacts of agricultural robotics on the farming industry and to identify the key requirements and challenges that must be addressed to realize the benefits of this technology. Through a review of the papers published in the last seven years, each section of this article attempts to provide an overview of the capabilities and limitations of agricultural robots and highlight the opportunities and obstacles that must be overcome in order to fully harness the potential of this innovative technology in the pursuit of more sustainable, efficient, and profitable farming practices. A review of the advances in robot manipulators, soft robotics (both in soft manipulators and soft grippers), and field robots is presented with an emphasis on the engineering solutions in their development. We also present an overview of existing autonomous field robots in arable farming, evaluate their suitability, and highlight the main challenges and requirements based on literature findings to give an outlook on possible future solutions. In this study, field robots are referred to as autonomous mobile platforms that perform farming tasks inside indoor or outdoor environments and are expected to navigate autonomously without disruption and avoid any obstacle placed within the confinement of movement. This study provides insights for farmers, policymakers, and technology developers looking to understand and seize the opportunities presented by agricultural robotics.

4.2 ADVANCES IN ROBOTIC MANIPULATORS FOR AGRICULTURE

Innovations in terms of robotic manipulator control in digital agriculture have advanced considerably in the last decade, with the aim of reducing costs and increasing efficiencies. The availability of compact imaging sensors, such as digital cameras that include machine learning and AI and can be trained to perceive depth information, besides the flexibility of open-source image processing software packages that have been customized for different applications have played significant roles in accelerating this sector. Preliminary studies have shown that the majority of available robot manipulators in agriculture are using Image-Based Visual Servo (IBVS) control to reach a target position. The presented study provides an overview of different redundant manipulators that are controlled by means of visual servoing for automating various field tasks in digital agriculture including (1) pruning, thinning, and trimming, (2) harvesting, and (3) inspection and target spraying. The reviewed works suggest that developing optimal tree shapes and planting techniques is necessary to improve the performance of visual servo control and automate farming operations with robots. In addition, selecting the right imaging sensors, employing graphics processing units, and training the computer vision algorithms with more fruit and plant datasets have been highlighted as the three main elements for improving the functionality of IBVS in manipulator control for agricultural applications.

Various published works have reviewed the latest achievements in the use of non-mobile robots for agricultural applications, including those that are used for autonomous weed control, field scouting, mowing, and harvesting [13]. Some of the active fields toward developing a robust and flexible robotic platform for farming applications include object identification, task planning algorithms, digitalization and optimization of sensors, synchronization of multi-robots, human-robot collaboration, and environment reconstruction from aerial images (or ground-based sensors) for the creation of virtual farms. More recently, trending research in robot manipulators control is towards building multi-robot arms that operate separately using distributed visual servo control systems and are installed on a mobile platform for optimizing a task such as fruit harvesting (Figure 4.1). An example includes a dual-arm robotic prototype system shown in Figure 4.1a that was designed for harvesting pears and apples on a joint V-shaped trellis and is claimed to be capable of picking fruits at the same speed as a human [33]. The robot uses an Intel Realsense D435 RGB-D (red, green, blue, and depth) camera, an infrared projector, and an adapted version of Mask R-CNN [34] that runs on a Jetson AGX Xavier processor for fruit detection [33]. Figure 4.1b shows the FFRobotics solution with multiple linear robot arms that combines accurate yet simple controls and rapid image processing with advanced algorithms for harvesting ripe apple fruits. These kinds of machines can be modified for harvesting different types of fresh fruits. Figure 4.1c shows the RB-VOGUI mobile robot platform that has been integrated with two Universal robot arms for active inspection and picking grapes in vineyards.

FIGURE 4.1 Multi-robot arms used for increasing efficiency: (a) dual-robot arm for pears and apples harvesting (image: NARO), (b) apple harvesting machine with multi-robotic arms (image: FFRobotics), and (c) integration of a mobile robot with dual-robot arms for grapes harvesting (image: Robotnik).

4.2.1 ADVANCES IN VISUAL SERVOING AND COMPUTER VISION

As a method to achieve object tracking and control the motion of a robot with the feedback extracted from a vision sensor, visual servoing is classified into two main classes: image-based visual servoing (IBVS) and position-based visual servoing (PBVS). The information to control the end effector of a robotic system is collected by one or more cameras observing the operation range. The main advantage of this technique in agricultural robotics is the increased flexibility for robots of all kinds in unstructured environments and dense orchards. The two configurations for positioning the camera in VS are eye-in-hand [23] and eye-to-hand [35]. In the first method, the camera is mounted on the manipulator or the gripper, providing a direct but limited view as it is closer to the target. This configuration is widely used for harvesting robotics in dense vegetation [23, 36], including automated harvesting of tomato [37], citrus [38], sweet pepper [23, 39], and cucumber [40, 41], since it provides the robot with the possibility of exploring the workspace, tracking fruits and branches, and interacting with the objects in its vicinity that require specific viewpoints. The main drawbacks with this configuration are the high computational requirements, synchronization between the camera and manipulator motion, and the need for additional mechanical supports to secure the camera on the robot. The second method involves placing the camera on a fixed platform separate from the robot manipulator to provide visual feedback for controlling the manipulator's movements. This configuration offers simplified calibration, reduced complexity, and higher flexibility for accurate object tracking, making it well-suited for improving the performance of eye-in-hand visual servoing tasks [36]. Due to the separation of the vision system from the manipulator, a panoramic sight of the workspace is possible using an extra camera with the eye-to-hand configuration, resulting in more fruits being detected with a more stable visual reference frame [42, 43].

The IBVS technique, which is based on the error between the current and the desired visual features, is shown in Figure 4.2. In order to reach the desired feature set, the error to the current feature set has to be minimized. Therefore, this method uses 2D image data in contrast to the PBVS technique, which is a model-based approach. Several research works have studied the possibility of cooperation among multiple cameras [38, 42]; for example, hybrid systems consisting of a fixed eye-to-hand camera and an eye-in-hand camera [44]. The control of both cameras is completely independent in this case, and the global camera controls the translating degrees of freedom with a landmark at the end of the translating joins. Moreover, the integration of artificial intelligence and computer vision techniques besides optimized manipulation strategies has demonstrated significant improvements in the accuracy and efficiency of the visual servoing [45, 46]. Although traditional computer vision techniques often rely on manually labelled features, which can be limiting in their ability to accurately classify complex agricultural objects or scenes, segmented deep learning [47] approaches employ convolutional neural networks (CNNs) [48] to automatically learn and extract features from segmented regions, enabling precise object recognition and localization [49]. Recently, Bayesian segmented deep learning [50, 51], which combines the advantages of both Bayesian inference [52] and segmented deep learning [53], has been employed to address the limitations of traditional deep learning techniques by introducing uncertainty estimates. This has resulted in more reliable predictions in complex agricultural scenes with occlusions, varying lighting conditions, and cluttered backgrounds [54].

It is a common practice to use the RGB camera sensor in visual servoing in combination with distance detection sensors, especially when the robot is subjected to work in environments such as an agricultural field in which the target can be occluded by the leaves. In this case, the feedback is provided by both the camera and the other sensors (i.e., infrared sensor), known as RGB-D cameras [56, 57]. This architectural choice is due to the difficulty to calculate the distance between the arm and the desired object only by a 2D image with a single in-hand camera. Furthermore, the visual servoing software routine is implemented on the low-performance application processor. RGB-D sensors have computer vision blocks that estimate the position of the desired object in the image as a pixel coordinate. Instead, the distance estimation block of an RGB-D camera calculates the distance in the z direction between the hand and the object. The visual servoing block of the robot is the main system controller and drives the arm using a set of motors. Figure 4.3 shows five of the most popular RGB-D cameras, including the Intel® Realsense [58], that are widely used with robot arm manipulators in agriculture.

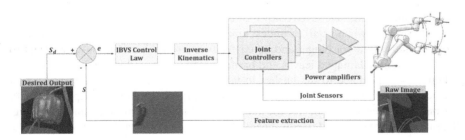

FIGURE 4.2 Image-based visual servoing with the eye-in-hand configuration for robotic harvesting [55].

FIGURE 4.3 Common RGB-D cameras used in visual servoing of agricultural robot manipulators.

Simulation environments such as Robot Operating System (ROS), MATLAB® (The MathWorks Inc, Natick, MA, USA), and CoppeliaSim (Coppelia Robotics AG, Zurich, Switzerland) provide designers with a flexible approach for experimenting and implementing IBVS algorithms in order to optimize plant/fruit localization, improve plant/fruit scanning, and develop strategies for finding collision-free paths [59]. Published studies demonstrate a completely simulated workspace environment, including a replica of the robot manipulator in Inverse Kinematic mode, object tracking, and the orchard or the field with bushes and plants [55, 59]. It should be noted that any visual servo system must be capable of tracking image features in a sequence of images. A simulation demo is presented in Figure 4.4 for sweet pepper harvesting based on the image moment method [60]. In an actual experiment with one or more RGB-D cameras, feature-based and correlation-based methods (as well as artificial intelligence and deep learning training methods) are used to improve the robot's image classification for tracking. These studies use image data taken from the robot camera as the input to the IBVS control algorithm in MATLAB. The extracted information is then fed back to the simulated workspace for IK calculation and for determining the trajectory path to the fruit. In most cases, ROS is used for bi-directional communication between the simulated environment and the robot via its publish and subscribe architecture. The proposed approach allows researchers to review, approve, and execute different trajectories for placing the sensor in the most desired positions, as well as providing a flexible framework for evaluating different sense-think-act scenarios to verify the functional performance of future manipulators with zero risk to the robots and operators.

4.2.2 Advances in Robotic Pruning, Thinning, and Trimming

Pruning refers to the selective removal of branches ahead of the spring growing season. Thinning is when small or undeveloped fruits, known as fruitlets, are removed from trees to allow for bigger and better fruits to grow. Trimming is carried out to cut the dead and damaged leaves and branches of bushes, and in some cases to give a better-looking shape to the bushes in terms of growing more fruit carrying into the next growing season. Figure 4.5 shows three different robotic manipulators with vision-based control systems for the automation of pruning, thinning, and trimming. These robots are expected to significantly contribute to cost savings by reducing the demand for the human workforce.

Reports indicated that pruning of apple trees comprises about 20% of total pre-harvest production costs [13]. Between 30–35 working hours of skilled labor are

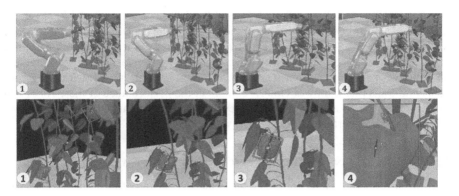

FIGURE 4.4 Simulation of visual servo control experiment with the eye-in-hand configuration and PID control law on joint angles with feedback from image moments. Stability was achieved in 2.5 seconds [55].

FIGURE 4.5 Vision-based control of robotic manipulators for automation of (a) pruning [61] (photo courtesy of Penn State University), (b) thinning [62] (photo courtesy of Ai-Ping Hu, Georgia Tech Research Institute), and (c) trimming [63, 64].

required per acre for the manual pruning of apple trees. Determining branch diameter is one of the most important and challenging parts of robotic pruning. Figure 4.5a shows a robotic pruner developed by Penn State University that combines a three-rotation wrist end-effector for cutting the branches, and a three-directional linear manipulator that houses and moves the end-effector to targeted pruning locations. The robot uses a light detection and ranging (LiDAR) sensor for the 3D reconstruction of apple trees and an RGB-Depth camera vision system to accurately measure the branch diameter to automatically make the pruning decisions. Figure 4.5b shows an intelligent thinning robot to take over the manually intensive tasks of thinning and pruning peach trees. This robot also uses a LiDAR sensing system and RTK-GPS for autonomous navigation within the orchards while avoiding collision with random obstacles. The detection of peaches is realized using an embedded 3D camera and a claw end-effector. Figure 4.5c shows the Trimbot robot that can navigate over different terrains, approaches boxwood plants, and trims them to the desired shape. The robot platform is based on a modified Bosch robot lawn mower, which navigates autonomously using 3D-based vision scene analysis. During trimming, a robotic arm is controlled by visual servo in order to trim the bush. A novel end-effector had to be designed to guarantee the flexibility of the manipulator, precision of trimming, and smoothness of the trimmed bush surface. It should be highlighted that the outdoor light conditions (i.e., sunny or cloudy) can affect the performance of these robots. In an orchard, no two bushes or trees are ever the same, hence the technology is not yet mature enough to perform these tasks as well as a human.

4.2.3 ADVANCES IN ROBOTIC WEEDING AND TARGET SPRAYING

Robotic weeding and target spraying contribute significantly to overall crop production and are the two main viable solutions for efficient weed removal and reduction of chemicals used in the fields that also contribute to lesser soil compaction. However, the development of these robots is challenging due to several specific complexities, including (1) correct identification between weeds and healthy plants, which can be extremely difficult due to the variation in plant physiology and genetic differences in size, shape, and color; (2) non-uniform locations of weeds that can be deteriorated more due to rains and wet muddy soil; (3) shading and occlusions; and (4) control of the robot arm under strong winds and volatile climate conditions. The two main approaches toward automated weeding and target spraying are the offline method (using georeferenced prescription maps) and the on-the-go method (also referred to as the real-time or online method, which uses weed detection sensors on the robot). Similar to the studies on robotic harvesting, RGB-D cameras are also used in robotic weeding, target spraying, and phenotyping, but other optical sensors are usually involved too. For example, Figure 4.6 shows three different robot arms, each with a different sensing solution that has been mounted onto commercial autonomous mobile platforms such as Husky A200 or Jackal for performing real-time weeding, spraying, and phenotyping. LiDAR solutions, such as the Velodyne's Puck VLP-16

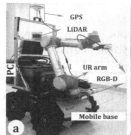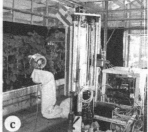

FIGURE 4.6 Examples of robot manipulators with different sensing systems used in weeding, target spraying, and phenotyping: (a) 6-DoF UR5 robot arm mounted on a Husky A200 mobile robot for automated weeding in cornfields [65], (b) 7-DoF Kinova Gen3 robot arm for target spraying of vineyards [66], and (c) CROPS manipulator for target spraying of grape vines [67].

shown in Figure 4.6a, are used for constructing a real-time 2D or 3D navigation map of the field at close range and for providing real-time 3D pointcloud information for precise navigation [65]. In general, these robots require higher-performance onboard computers, such as the workstation PC shown in Figure 4.6a, that can carry out rapid computation to simultaneously detect the weed and control the robot arm. As a result, the operating speeds of most of the available prototypes are still below the threshold level to be deployed in large-scale farms. For the case of spraying robots, the small size of the robot platform and the spraying tanks is also a limitation, suggesting that these robots should be mounted on autonomous tractors with large chemical tanks.

The weeding robot shown in Figure 4.6a uses the RealSense D435i depth camera for multi-target depth ranging and a quadratic traversal algorithm for shortest path planning [65]. To recognize weeds and corn, a fast R-CNN neural network is implemented for fast real-time recognition. The experimental results with this robot have shown a success rate of 90.0%. Figure 4.6b shows the Kinova Gen3 7-DoF robot arm that has been integrated for target spraying [66] and uses Optitrack cameras to measure the position of the spray frame on-the-go. Optitrack markers are attached to the end-effector of the robot arm. The robot is still under development and is planned to be based on a combination of deep learning and depth information captured by an RGB-D camera. The robot arm uses an Ethernet connection to communicate with ROS drivers running on the PC. Low-level control of the arm is then achieved via joint velocity commands, while encoder measurements from arms are used as feedback. Figure 4.6c shows a custom-built robotic system with eye-to-hand visual servoing configuration for selective targeting of pesticide applications that deposit chemical spray only where and when they are needed and at the correct dose. The sensing instruments for this robot are installed on a frame beside the manipulator with a 700 mm longitudinal offset between the camera and the robot arm base. The robot benefits from disease-sensing sensors by means of multispectral imaging of a grapevine canopy under diffuse illumination. Multispectral images of the canopy are acquired by a 3-CCD, R-G-NIR camera model MS4100 (DuncanTech, Auburn, CA, USA), which captures 1912- by 1076-pixel, 8-bit images in three distinct spectral channels: green (540 nm), red (660 nm), and NIR (800 nm). The multispectral camera and a regular RGB camera are mounted on a sliding holder that allows positioning adjustments from 800 mm to 2500 m in height. A computer is also mounted on the robot base and is dedicated to fast embedded image processing of multispectral images for disease detection, target position computation, and transmission to the manipulator control computer.

4.2.4 ADVANCES IN ROBOTIC HARVESTING

In general, a robot has to perform five main tasks for harvesting, including fruit detection, fruit localization, image processing and extracting information from cameras, inverse kinematics, and path planning. The main challenges in robotic harvesting have been defined as the detection and localization of fruits under high-density leaves and the changing outdoor light intensity [13, 68]. Recently, RGB images have been used in combination with deep learning algorithms to overcome the issues with shadowing and occlusion [68, 69]. To avoid collision between the robot arm and the tree branches or the plant supporting system, path planning is performed before the robot manipulator is controlled in inverse kinematic mode. A review of the published literature reveals that most of the studies on robotic harvesting have used commercially available UR3 or UR5 robot manipulators from Universal Robots due to their compact size, light weight, accuracy, and easy control. Virtual replicas of these arms are available in various robot simulators and can be used for experimenting with different harvesting scenarios. For the vision-based control of the robot, RGB-D cameras that can simultaneously obtain RGB and depth information are used to detect and locate the fruits in the images. Figure 4.7 shows four robotic harvesting manipulators that benefit from RGB-D cameras in their sensing and control. Multiple studies have reported that the position of the pointcloud that RGB-D cameras acquire under unsteady outdoor light conditions is inaccurate [33, 42, 70]. In addition, fruits need to be judged for ripeness and harvesting, which can be extremely challenging to determine solely from color information.

Figure 4.7a shows a dual-arm fruit harvesting robot [68] that consists of four Real sense D435 RGB-D cameras for detection and localization of the fruits and two robot arms (UR3 and UR5) to increase work efficiency by harvesting the upper side and lower side of the trees simultaneously. The robot has a central computer that performs all of the computation and control tasks. Two of the robot cameras are adjusted to look up at the fruit tree from directly below, one to look diagonally upward, and one to look directly to the side. This configuration allows viewing the fruit tree from different directions and minimizes the number of fruits in the blind spots hidden behind leaves and branches. Deep learning is performed on the RGB image acquired from the RGB-D camera to detect the position of fruits in the image. Next, the 3D positions of the fruits are identified by combining the positions of the fruits in the RGB image and the depth image. To this aim, the RGB images are acquired from the RGB-D cameras mounted on the robot, and the fruits in the images are detected. It is necessary to combine information such as color and texture in order to achieve sufficient accuracy. Single Shot Multibox Detector (SSD), which is one of the object detection algorithms, is applied to detect fruits in images. SSD is a method for detecting objects in images using a single neural network. The reason for using SSD was due to the speed and accuracy of the method. All of the information about the detected bounding box D is obtained from the results of the detection of the fruit in the image by SSD. Figure 4.7b shows the Harvey sweet pepper harvesting robot [69], which has achieved a 76.5% success rate on 68 fruit within a modified scenario.

FIGURE 4.7 Image-based visual servo control with eye-in-hand configuration using RGB-D camera for (a) dual-arm fruit harvesting [68], (b) 6-DOF UR5 robot arm for sweet pepper harvesting [69], (c) 3-DOF custom-built manipulator for apple harvesting [70], and (d) 6-DOF UR5 robot arm for apple harvesting [71].

Harvey benefits from a 6-DOF UR5 for manipulation and an RGB-D camera as input for the perception system (segmentation of sweet peppers and peduncles). During the fruit segmentation stage, the robot captures a 3D color image of the whole scene using an eye-in-hand RGB-D camera. A target sweet pepper is localized at the long-range perspective, and its image is used as an input to the control system to move the camera to a close-range perspective of the targeted fruit to improve the performance of the peduncle segmentation. The next stage localizes the peduncle of the target sweet pepper using a Deep Convolutional Neural Network and a 3D filtering method to estimate the centroid of the peduncle. A grasp selection is then performed on the segmented 3D points of the targeted sweet pepper. The grasp selection uses a heuristic to rank possible grasp poses on the target sweet pepper using surface and position metrics. Figure 4.7c and Figure 4.7d show two robotic platforms for apple harvesting that also benefit from RGB-D cameras and IBVS [70, 71]. These two studies suggest that the development of an affordable and efficient harvesting robot requires modification of the cultivation systems for overcoming the problems of fruit visibility and accessibility. Robotic harvesting must be economically viable, which means it must sense fast, calculate fast, and move fast to pick a large number of fruits every hour that are bruise-free. Table 4.1 summarizes different control approaches and cameras/sensors used in the research and development of robotic harvesting for various crops.

TABLE 4.1
SUMMARY OF REVIEWED HARVESTING ROBOTS FOR IDENTIFYING CONTROL STRATEGIES AND CAMERA/SENSORS

Crop	Manipulator	DOF	Controller	Camera/Sensor	Ref
Sweet Pepper	Fanuc LR Mate 200iD	6	PID	Logitech C920 HD Pro USB/ proximity Hokuyo URG04LXUG01	[55]
Sweet Pepper	Baxter	7	ROS *actionlib* service	USB CMOS color Autofocus Camera (DFK 72AUC02-F, TheImagingSource)	[23]
Sweet Pepper	CROPS	9	IK-based	VRmMS-12; CamBoard nano/SR4000; Prossilica GC2450C	[39]
Apple	Multi-joint vertical manipulator	4	Machine Vision Feedback P-based	color CCD camera/laser ranging sensor	[72]
Apple	PRRRP Structure manipulator	5	IBVS	color CCD camera/infrared double photoelectric cells	[73]
Apple	UR3	6	IK-based	ZED	[74]
Citrus	Robotic Research K1207 manipulator	7	Decoupled PD-based	color CCD camera (KT&C, KPCS20-CP1)/infrared proximity sensor	[75]
Aubergine	Dual-arm Kinova MICO™	6	PID control	Prosilica GC2450C; Mesa SwissRanger SR4000	[76]

4.3   HUMAN-ROBOT COLLABORATION

Major factors that are limiting the widespread use of robotics arms in agriculture are related to effectiveness, costs, and issues with electrical power consumption (battery) in actual field conditions. Most of the published studies in this area conclude that to this date the best existing visual servoing algorithms, regardless of the manipulator, are unable to continuously and efficiently perform tasks such as pruning, harvesting, weeding, or target spraying in order to completely replace human workforces and meet the farmers' requirements [32]. Therefore, teleoperation [77] or human-robot collaboration [78] might be necessary to solve the challenges in robotic arm control for enhancing their functionalities. For example, in collaborative harvesting with a human-robot interface, occluded fruits that are missed by the robot vision system are identified by an operator on a screen using input devices such as a mouse or Sony PS3 Gamepad (as shown in Figure 4.8), or the robot actuation can be controlled entirely in a virtual environment. The main limitation of these solutions is the communication between the human and the robot for exchanging live-streaming images and control messages, which are currently realized via WiFi [79].

Among these technological advancements, collaborative robots, commonly known as Cobots [80, 81], have emerged as a promising solution to work safely alongside humans in a shared workspace. While Cobots bring a new level of efficiency and productivity to agricultural operations, their initial investment costs may prevent them from being viable for small-scale farmers. Studies assessing their economic performance in comparison to conventional labor for four different cultivars in Greece in a lifecycle costing methodological framework have shown that the annual equivalent costs could be reduced by up to 11.53% using Cobots [82, 83]. The same study indicates the great potential of Cobots regarding specific operations such as weed control, pruning, herbiciding, and topping, in order to cope with climate change impacts and excessive energy consumption. A review article in this field has concluded that the use of Cobots is not yet widespread, but is still open to further development in order to reach the stage of commercial availability in the near future [80]. Some studies suggest applying a framework for switching the levels of autonomy in human-robot collaboration research that determines whether the current robot operation is fully autonomous, semi-autonomous, or teleoperated (manually

FIGURE 4.8   Human-robot collaboration for increasing the efficiency of robot control in agriculture: (a) teleoperation via simulation environment (image: Adaptive AgroTech), (b) selecting target grape clusters using a mouse, (c) digital pen on a smart interactive touchscreen whiteboard, and (d) Wii remote [79].

controlled). Berenstein et al. [84] describe and characterize a collaborative human-robot framework for site-specific spraying of grape cluster targets. The human-robot framework is designed such that the human assists the robot with target detection, and the robot detects targets and performs the spraying. Four different collaboration levels were tested, and the number of true positive and true negative detected targets were analyzed. According to the results, the human-robot collaboration successfully performed the complex spraying tasks, and a reduction of spraying material of 50% was reached. In case of a high probability of dangerous pest occurrence, a maximum of true positive detected targets is requested, which means a full manual collaboration is recommended. In case of a low probability of dangerous pests, a minimum of false positive detected targets is preferred, that is, a robot makes a decision and a human supervisor approves that decision. According to E. Aivazidou et al. [84], social acceptance barriers are underestimated concerning the automation of tasks that are currently still performed by field workers. Therefore, human-robot synergy systems are a promising, socially viable alternative for sustainable rural development. Additional studies for the possibility of using 5G and LoRaWAN are underway to remotely control the robotic arm more stably.

4.4 ADVANCES IN SOFT ROBOTICS AND SOFT GRIPPERS

The complex nature of the agricultural fields and the required flexibility of the manipulator tasks are the main motivation to replace expensive industrial robots that have rigid components and heavy actuation mechanisms with soft robotic arms and grippers. Soft robotics, particularly the field involved in the development of robotic grippers and manipulators, has emerged as a solution to complex manipulation tasks in medical and industrial fields [85–88] and is undergoing research for agricultural applications [31, 89, 90]. Attempts to emulate human skills in terms of grasping ability during agricultural tasks, such as harvesting, have resulted in numerous mechanical end-effectors and robotic arms. With the emergence of soft robotics, grippers and robotic manipulators based on soft, deformable materials have recently begun to be proposed for agriculture [90, 91], as one of the fields where soft robotics can offer the greatest benefits are tasks related to high-value crops. An example includes the work of Chowdhary et al. [92], in which a small and low-cost multi-purpose agricultural mobile platform with a soft robot manipulator (Figure 4.9) has been proposed, demonstrating the potential to be used in dense berry orchards for complex tasks, including scanning berries for ripeness, harvesting, and health assessment. Figure 4.9a shows the proposed architecture of the hybrid mobile robot platform with a continuum arm that is mounted on a rigid link, and Figure 4.9b presented preliminary experiments with the concept.

4.4.1 Soft Robotic Manipulators

In contrast to rigid or classical robotic manipulators, soft robotics manipulators have the advantage of being able to interact with humans, have a low cost, and are well-suited to work in unstructured environments such as in the agricultural sector. They can perform tasks such as crop monitoring for optimal fruit ripening point selection by means of flexible movement of a camera, harvesting, controlling weeds, detecting insects and diseases throughout the dense plant canopy, and pruning and thinning branches [92]. However, challenges remain to be addressed in terms of control strategies, as their complexity lies in their inherent flexibility in the joints, their virtually infinite degrees of freedom, and the nonlinearity of the materials used [93]. Figure 4.10 shows different actuation techniques used to drive soft robotic arms. In both Figure 4.10a and Figure 4.10b, the arms are pneumatically actuated, while in Figure 4.10c, the arm is cable-operated.

Several approaches for measuring the characteristics of soft actuators have been proposed in the literature. In one study [96], a spherical object connected to a force sensor via an inextensible cable is grasped by the soft actuator mounted on a motorized platform to measure slip properties. Another study [97] used a six-axis force transducer for a similar approach. Alternatively, pressure-mapping sensors were employed [98] to measure contact force and pressure, providing a reliable measurement for grasping static objects. Grip strength is measured in a similar way as in the previous studies. In [99], a payload test was conducted to obtain the grip strength, while FEM software was used to determine the contact pressure, although

FIGURE 4.9 Fundamental research in the use of soft robotics for agriculture: (a) a proposed agricultural robot platform with a continuum arm [92], and (b) preliminary experiments with a low-cost and scalable soft robotic as a promising solution for robotic harvesting [92].

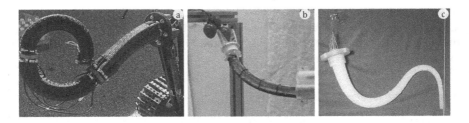

FIGURE 4.10 Soft robotics manipulators: (a) Octoarm, a soft robotic manipulator [94], (b) soft manipulator for agroforestry tasks [92], and (c) tendon-driven soft robotic manipulator [95].

inaccuracies may arise due to its dependence on the mathematical model of the material used. Lastly, a detailed analysis of parameters such as grabbing height, pressure, and motion acceleration was proposed in [100], including tests in both static and dynamic conditions, vertical and horizontal positions, and taking into account variables such as size, weight, constituent material, actuation pressure, and grabbing height. One of the main contributions of this study is the introduction of the handling ratio, which offers a measurable performance comparison.

4.4.2 INNOVATIONS IN SOFT GRIPPERS

Soft grippers are used to produce a significant advance in the manipulation of delicate objects. In the industrial sector, companies such as Soft Robotics Inc. or Qbrobotics are already providing soft gripper-based solutions. The main asset of this technology is the possibility of continuously varying its shape without the need for complex multi-joint mechanisms while presenting lower costs and simpler structures than hard end-effectors. However, the implementation of this solution in the agricultural sector is still a challenge. Intrinsic characteristics such as adaptation to unstructured environments, grip without bruising fruit, and the ability to work alongside humans make agriculture a potential sector for the implementation of this technology. Figure 4.11 shows three different soft grippers designed for harvesting tasks.

Figure 4.11a shows a gripper of fluidic elastomer actuators (FEAs) type. These classes of soft grippers are distinguished by their ability to manipulate large and heavy objects. Such grippers can be relatively demanding in power consumption compared to other soft technologies due to the pneumatic system [90]. In Figure 4.11b, a passive or topology-optimized gripper can be seen. The design of these grippers, which are driven by a servomotor, is mechanically optimized to manipulate a particular object, which ensures undamaged manipulation. Since it is custom designed, its main limitation is that it is only suitable for picking a target fruit. A technology widely used for its precision in position control is the so-called tendon-driven gripper, shown in Figure 4.11c, in which the main challenge is the complexity of maintaining tendon tension and avoiding its fracture [104]. The schematic shown in Figure 4.12 is an example of the electrical and pneumatic circuit scheme for the control of a pneumatic soft gripper with an opening and closing movement.

FIGURE 4.11 Harvesting soft grippers: (a) pneumatic soft gripper for apple harvesting [101], (b) passive structure soft gripper for apple harvesting [102], (c) tendon-driven soft gripper for blackberry harvesting [103].

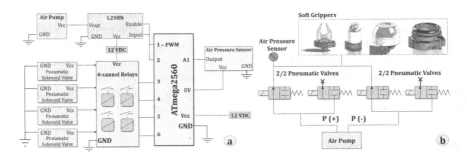

FIGURE 4.12 Schematic presentation example of the (a) electrical and (b) pneumatic circuits for controlling a soft gripper.

In order to validate the feasibility of the proposed approaches and evaluate their performance, several experimental tests are commonly performed, including a grasping force test, in which the gripping force is evaluated. Figure 4.13 shows a test example of a soft gripper on a robotic manipulator. Figure 4.13a shows the gripper is in its extended mode with the position that is used to adapt to the various sizes of the fruit. Figure 4.13b shows the gripper in its closed mode, where it can be used to move leaves or branches. In Figure 4.13c, another more precise type of grasping can be seen, where in addition to serving for picking, it also serves for positioning the fruit in pick-and-place tasks as well as for putting aside branches or leaves for easier picking in a possible manipulation situation. Finally, Figure 4.13d shows the position of the gripper, where it has more pulling force to pick the fruit since the object is completely blocked.

4.4.3 Finger-Tracking Gloves

The use of finger-tracking gloves is a novel approach to the study of fruit picking. Traditionally, the visual method has been the most common approach in the scientific literature to identify harvesting movements. This method, although simple, is inaccurate to measure in detail the motion pattern made by humans in fruit harvesting. With the use of finger-tracking gloves (see Figure 4.14), it is possible to monitor numerically the movement patterns performed during agricultural tasks, such as harvesting. This allows for analyzing the different movements involved in the harvesting, also known in the literature as picking patterns [90], in a detailed manipulation study. One of the key studies that has contributed to the introduction and development of finger-tracking gloves is the work of Beyaz (2018) [105], in which a prototype harvest glove with an affordable electronic system was successfully used to determine mechanical damage on apples. The system demonstrated proof-of-concept and was found to be capable of measuring impact forces and generating warning signals.

An application of finger-tracking gloves is in analyzing blueberry picking patterns. Figure 4.15 shows the results of a manipulation study, where the Manus Prime 2 finger-tracking gloves were used. These types of gloves can track the angles between

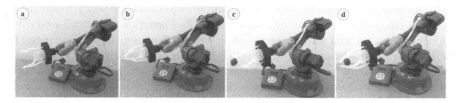

FIGURE 4.13 Testing a soft gripper on the WLkata Mirobot robotic arm showing (a) extended position, (b) closed position, (c) blueberry grip with the tip, and (d) blueberry grip in the locked position for harvesting.

FIGURE 4.14 Blueberries harvesting with Manus Prime II finger-tracking gloves.

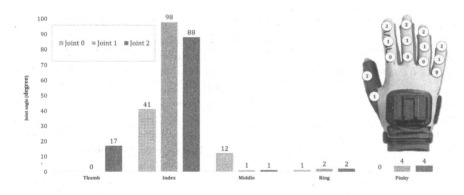

FIGURE 4.15 Maximum angle reached by the finger joints during blueberry harvesting.

the different joints as well as the angle of the stretch between the fingers. As can be seen, the thumb and index finger were used in most of the manipulations. The spread angles of the thumb, index, middle, ring, and pinky were 39°, 0°, 0°, 0°, and 0°, respectively. It is important to mention that those spread angles did not vary, particularly the angle of the thumb and index finger remaining completely static, which explains a greater stiffness in the grasping motion.

In summary, soft robotics, in particular soft manipulators and soft grippers, is an ideal technology for the agricultural sector due to its adaptation to unstructured environments, low cost, and gentle crop handling without damaging or bruising them. In the study of manipulation in agricultural tasks, tools such as finger-tracking gloves can be used not only for the design of the suitable soft gripper for each application but also for the detailed analysis of the picking patterns or as input for path planning for the robotic manipulators where these end-effectors will be attached.

4.5   ADVANCES IN FIELD ROBOTS AND THEIR AVAILABILITY IN EUROPE

One of the promising solutions for increasing the profitability of farms is the more efficient use of inputs. Field robots reduce the workload of the farmers for repetitive tasks such as mechanical weeding, mowing, and spraying, and can contribute to higher yields and quality through site-specific applications of fertilizers and pesticides. Employing robots for these tasks has a high potential for saving costs, but this is justified if additional cultivation steps or a higher repetition rate leads to a qualitative or quantitative increase in yield [6]. A recent study on the acceptance level of agricultural robots in Germany clearly indicates that the majority of the farmers surveyed are keen to immediately use this technology on their farms, especially for tasks such as weeding, due to the potential benefits of saving labor and practicing more sustainable farming methods [106]. A survey in the published studies also shows that small robots are particularly of interest in small and irregularly shaped fields, where large machinery are unable to operate efficiently. The scalability of field robots and suitability for small field sizes besides their lower ownership costs create opportunities for smaller farms to become economically viable [107].

In order to identify the market availability and status of development (SoD) of field robots, a comprehensive search was carried out through related websites and published research works with regard to those robots that are commercially available (CA), robots that are under field tests and trials (UFT), and those that are still in prototype development (PD). Robots specialized in certain types of cultivation, such as viticulture, greenhouses, or orchards, are not included in this study, as both the plant habitus and the cultivation methods are very different compared to field crops. In this chapter, field robots are defined as ground autonomous vehicles, and our analysis is limited to products that are available on the European market due to various current trade restrictions. The analysis considered publications of robotic systems until October 2022 using the specified keywords, which revealed 35 available field robots on the European market, as summarized in Table 4.2. It should be noted that this review does not claim to be exhaustive, due to the dynamic character of this sector.

The determined robotic systems are divided into three categories, including 18 field robots that are available on the market, 4 robots that are in the test phase and in field trials, and 13 robots that are available as prototypes. On the scale of body weight, field robots can be categorized as lightweight (less than 300 kg), medium weight (300–1000 kg), and heavyweight (more than 1000 kg). These robots are equipped with one or multiple sensors for autonomous navigation (including collision avoidance), object detection, and feature extraction. Some of them have a flexible design that can be integrated with custom-built attachments to perform specific tasks such as mowing [108], weeding [109], and spraying [110]. Moreover, those that are equipped with several data acquisition devices, such as multispectral [111], hyperspectral [112], NDVI [113], thermal [114], or NIR cameras [115], provide a great opportunity for field scouting [116], early disease detection [117], and yield estimation [118], [119].

These robots are required to withstand harsh field conditions, have a flexible control design with interchangeable and compatible components, and benefit from a reliable navigation system with collision avoidance capabilities [120]. The last feature is a fundamental requisite for autonomous robots, with the sensors playing the most important role, as they are expected to provide accurate feedback to the controller. Studies have shown the feasibility of using a distributed collision avoidance system with multiple laser range finders and multi-channel infrared sensor arrays as an alternative for high-end 3D LiDAR for assisted navigation of a four-wheel steering field robot in a GPS denial environment [121]. From the farmers' perspective, they prefer that depending on the task requirements, different modules such as sensors, actuating devices, and manipulators on a multi-purpose field robot can be easily swapped. If successfully integrated and implemented, these robots can play a key role in reducing agricultural production costs by decreasing the number of human workforces that are currently engaged in performing repetitive tasks.

4.6 A CASE STUDY FOR POTATO FIELDS

A field robot is subjected to a high degree of variance with respect to different environmental factors and cultivation practices. This section reports on a case study that was carried out to identify the key challenges and requirements of field robots for potato cultivation, and to recommend the most suitable robot candidate that is available in the market and can operate in potato fields. To this aim, a literature review was carried out by taking into account the development phases and steps in potato cultivation. The study was then supplemented by observations and random measurements executed on a test field during the 2022 season. For this purpose, potato plant height and ridge geometry were of particular interest. The test site was located in Schöneiche in the district Dahme-Spreewald (51°56'N, 13°29'E, 122 m above sea level) in the state of Brandenburg, Germany (Figure 4.16). The soil in this area is attributed to the soil class of slightly silty sand [106]. The field size was 7.4 hectares, and various potato varieties were grown during the 2022 season. The detailed description and argumentation for the use of mobile robots in potato cultivation is available in [144], where the requirements for the field robots to assist farmers with the complex steps of plant care, such as maintaining the ridge structure and mechanically regulating weeds and pests, have been discussed.

Standardized potato cultivators normally form ridges with a spacing of 0.75 m and a height of up to 0.35 m [145]. Thus, the track width of the robot should match the intra-row spacing (0.75 m) or a multiple thereof. The ridge height of up to 0.35 m necessitates a minimum ground clearance of at least 0.35 m. Fulfilling these two dimensions avoids damaging the ridge structure when driving through the potato field. The evaluation revealed that 9 of the 35 available field robots presented in Table 4.2 meet the geometric requirements derived from the potato cultivation steps. The task of autonomous weeding for potato fields can be carried out either by integrated tools

TABLE 4.2

Summary of the Identified Field Robots in the European Market and Their Status of Development (SoD). CA: Commercially Available, UFT: Under Field Test, and PD: Prototype Development

Robot Image	Robot Name	Weight (kg)	Sensors	Motor Type/ Power (kW/hp)	Lifting Capacity	Function and Application Area	SoD
	AgroIntelli: Robotti LR [122]	3000	LIDAR, GNSS-based virtual fencing, RTK-GPS positioning system (2 cm accuracy), Computer vision, Bosch Rextroth Load-Sensing	1 Stage V Kubota diesel engine 54/72 (kW/hp)	1200	4-wheel drive, autonomous tool carrier for standard implements for weeding, seeding, spraying, ridging. Does not require power take-off. (Image: www.agrointelli.com)	CA
	AgroIntelli: Robotti 150D [123]	3100	N.A	2 Stage V Kubota diesel engine 106/144 (kW/hp)	750	4-wheel drive, autonomous tool carrier for standard implements for seeding, weeding, spraying, ridging, mowing, rotovating, organic and conventional vegetables, including leeks, brassicas, onions, red beet, potatoes, fodder beet, sugar beet, and cereals. (Image: www.agrointelli.com)	CA
	Aigro: Up	100	Dual RTK GPS, Proximity sensors	Electrical motors powered by Dual Li-Ion battery	N.A	2-wheel drive, lightweight autonomous tool carrier for tasks such as weeding and mowing up to 5 to 10 hectare fields with one battery charge. The robot can be also controlled via phone and tablet. (Image source: www.aigro.nl)	CA

TABLE 4.2 (Continued)
Summary of the Identified Field Robots in the European Market and Their Status of Development (SoD). CA: Commercially Available, UFT: Under Field Test, and PD: Prototype Development

Robot Image	Robot Name	Weight (kg)	Sensors	Motor Type/ Power (kW/hp)	Lifting Capacity	Function and Application Area	SoD
	Kilter: AX-1 [109]	260	High precision GPS, vision-based high-speed	Hybrid vehicle, 48 V DC and a four-stroke generator for power backup	N.A	2-wheel drive, ultra-high-precision weeding robot (based on a modular design) for vegetables, including carrots, parsley root, spinach, radish, rocket, baby leaves, and celeriac. It applies micro droplets of herbicide. (Image source: www.kiltersystems.com)	CA
	Exobotic: LAND-A1	180	N.A	Electrical motors 24V (40 Ah) LiMnO2 2*1.15 (Nm)	50	2-wheel drive, for lighter agricultural use and tasks such as crop monitoring, weed detection/mitigation and precision spraying pm uneven terrain, with differential suspension system, Gigabit Ethernet port, Wifi 802.11n (100 Mbps) or 802.11ac (200 Mbps), Remote Control optional 1.5km LoS, redundant power supply inputs and automatic failover. (Image: www.exobotic.com)	CA

(Continued)

TABLE 4.2 (Continued)

Summary of the Identified Field Robots in the European Market and Their Status of Development (SoD). CA: Commercially Available, UFT: Under Field Test, and PD: Prototype Development

Robot Image	Robot Name	Weight (kg)	Sensors	Motor Type/ Power (kW/hp)	Lifting Capacity	Function and Application Area	SoD
	Exobotic: LAND-A3	1100	N.A	Diesel + hydraulic drive, commercially available, fully electric version available on demand, 20 kWh basic version, extendable to 40 kWh	600	4-wheel drive, a multipurpose versatile tool carrier to perform different agricultural tasks, with PTO hydraulics, Gigabit Ethernet port, Redundant power supply inputs and automatic failover. Up to three category 1 task-specific implements can be attached to the robot at the front, middle, and rear at the same time. This makes the machine extremely versatile. (Image: www.exobotic.com)	CA
	Sitia: Trektor [124]	2930	GPS RTK, Various redundant sensors	Hybrid motorization	N.A	2-wheel drive, the first hybrid robot to work on different crops including viticulture for both narrow and wide rows vineyards, market gardening, vegetable field crops and arboriculture, soil cultivation, spraying, and hoeing. Remote maintenance. (Image source: www.sitia.fr)	CA

TABLE 4.2 (Continued)
Summary of the Identified Field Robots in the European Market and Their Status of Development (SoD). CA: Commercially Available, UFT: Under Field Test, and PD: Prototype Development

Robot Image	Robot Name	Weight (kg)	Sensors	Motor Type/ Power (kW/hp)	Lifting Capacity	Function and Application Area	SoD
	Farmdroid: FD20 [125]	900	N.A	Solar-powered electrical motors	140	Weeding, seeding, sowing, sugar beets, beetroots, onions, spinach, rapeseed, and different herbs. No hydraulics. (Image: www.farmdroid.dk)	CA
	Farming Revolution: Farming GT [109]	900	N.A	Electrical motors 4×18 (kW), 8104.7 (Nm), 5 (kWh)	700	4-wheel drive, autonomous mechanical weeding and hoeing for cabbage, lettuce varieties, onions, corn, sugar beet, pumpkin, field bean, potato, canola, soybean, and wheat. No hydraulics. (Image: www.farming-revolution.com)	CA
	Pixel Farming Robotics: Robot One [126]	1100	Dual RTK-GPS positioning system (2 cm accuracy), 4 Stereoscopic depth cameras,	Solar-powered (3 panels) electrical motors 500 (Nm) per wheel, 10*1.020 (kWh)	N.A	4-wheel drive, plants and weeds control, non-inverting tillage (NINV), More than 40 types of crops, including green beans, carrots, beets, and lupine. Nvidia Jetson Xavier, 8-core 64 bit (GPU), 4G connectivity (100 Mbps). (Image: www. pixelfarmingrobotics.com)	CA

(Continued)

TABLE 4.2 (Continued)

Summary of the Identified Field Robots in the European Market and Their Status of Development (SoD). CA: Commercially Available, UFT: Under Field Test, and PD: Prototype Development

Robot Image	Robot Name	Weight (kg)	Sensors	Motor Type/ Power (kW/hp)	Lifting Capacity	Function and Application Area	SoD
	Agxeed: AgBot [127]	3200	RTK-GPS (2.5 cm accuracy),	2.9 l Stage 5 Deutz diesel engine, optional electric driven PTO 55/75 (kW/hp), 300 (Nm)	4000	Autonomous tractor for light tillage, seeding, weeding, mowing, swathing, tedding, and hoeing for vegetable farming and special crops (e.g., potato). Communication modules for bidirectional data transfer and RTK correction. (Image: www.agxeed.com)	CA
	Carré: Anatis Co-Bot [128]	1000	TRIMBLE GPS and CLAAS camera guidance	Electrical motors	N.A	4-wheel drive, crop scouting, hoeing, monitoring, processing of key indicators, and decision support. Greenhouses, red beet, sugar beet, sweet corn, cabbage, lettuce, forest nurseries, orchard, production of vegetables and ornamental plants. (Image: www.carre.fr)	CA
	Elatec: E-Tract [129]	850	N.A	Electrical motors	400	2-wheel drive, hoeing, sowing, mulching, spring tine harrow, hilling. Vegetables cultivation and viticulture. (Image: www.elatec.fr)	CA

TABLE 4.2 (Continued)

Summary of the Identified Field Robots in the European Market and Their Status of Development (SoD). CA: Commercially Available, UFT: Under Field Test, and PD: Prototype Development

Robot Image	Robot Name	Weight (kg)	Sensors	Motor Type/ Power (kW/hp)	Lifting Capacity	Function and Application Area	SoD
	Naïo Technologie: Oz [130]	150	RTK-GPS	Electrical motors	300	4-wheel drive, hoeing, weeding, making furrows, seeding, assist, transporting, small-scale vegetable farms. (Image: www.naio-technologies.com)	CA
	Naïo Technologie: Dino [131]	1250	RTK-GPS (2 cm accuracy), camera guided implement	Electrical motors	N.A	Weeding, lettuce, onions, carrots, parsnips, cabbage, leeks, cauliflower, various herbs (garlic, cilantro, mint, etc). (Image: www.naio-technologies.com)	CA
	Naïo Technologie: Orio [132]	1450	RTK-GPS (2 cm accuracy), LIDARs, geo fencing module	Electrical motors 4×3 (kW), 21.5 and 32.3 (kWh)	N.A	4-wheel drive, autonomous tool carrier, weeding lettuce, onions, carrots, parsnips, cabbage, leeks, cauliflower, various herbs (garlic, cilantro, mint, etc), row crops and beds of vegetables, arable crops, large growers and contractors. (Image: www.naio-technologies.com)	CA

(Continued)

TABLE 4.2 (Continued)

Summary of the Identified Field Robots in the European Market and Their Status of Development (SoD). CA: Commercially Available, UFT: Under Field Test, and PD: Prototype Development

Robot Image	Robot Name	Weight (kg)	Sensors	Motor Type/ Power (kW/hp)	Lifting Capacity	Function and Application Area	SoD
	Naïo Technologie: Jo [16]	850	RTK-GPS (2 cm accuracy),LIDARs, geo fencing module	Electrical motors 2*3 (kW), 16 and 21 (kWh)	250	2-wheel drive, Crawler robot carrying usual soil cultivation implements, vineyards. (Image: www.naio-technologies. com)	CA
	Small Robot Company: Tom [133]	349	N.A	Electrical motors	N.A	4-wheel drive, crop and soil monitoring: plant count, weed detection, spraying, arable crops, winter wheat. (Image: www. smallrobotcompany.com)	CA
	Small Robot Company: Dick [133]	N.A	N.A	Electrical motors	N.A	4-wheel drive, precision spraying, laser weeding, arable crops, winter wheat. (Image: www. smallrobotcompany.com)	UFT
	Dahlia Robotics: Dahlia 3.3–4.3	N.A	N.A	N.A	N.A	Weeding, sugar beets, and salads (Image: www.dahliarobotics. com)	UFT

TABLE 4.2 (Continued)
Summary of the Identified Field Robots in the European Market and Their Status of Development (SoD). CA: Commercially Available, UFT: Under Field Test, and PD: Prototype Development

Robot Image	Robot Name	Weight (kg)	Sensors	Motor Type/ Power (kW/hp)	Lifting Capacity	Function and Application Area	SoD
	Naiture GmbH: 8-Spur Jätroboter [134]	N.A	N.A	N.A	N.A	Weeding, carrots, beetroot, spinach (Image: www.naiture.org)	UFT
	Zauberzeug: Field Friend	100	N.A	Solar-powered electrical motors	N.A	Weeding, monitoring, harvesting (Image: www.zauberzeug.com)	UFT
	A.I.Land GmbH: ETAROB [135]	N.A	N.A	N.A	N.A	Weeding, fertilizing, harvesting, strawberries, potato, cauliflower, iceberg lettuce, celery (Image: www.a-i.land)	PD

(Continued)

TABLE 4.2 (Continued)

Summary of the Identified Field Robots in the European Market and Their Status of Development (SoD). CA: Commercially Available, UFT: Under Field Test, and PD: Prototype Development

Robot Image	Robot Name	Weight (kg)	Sensors	Motor Type/ Power (kW/hp)	Lifting Capacity	Function and Application Area	SoD
	BOSCH Deepfield Robotics: BoniRob [136]	1000	RTK-DGPS system, 2D- and 3D-laserscanner, Gyroscope, spectral imaging, 3D time-of-flight cameras	Hub motor	N.A	4-wheel drive, targeted weed control by phenotyping, tamping in or precise weeding, spraying of liquid fertilizer or crop protection agent of individual plants. Over time, the Bonirob learns to better distinguish between desired and undesired plants based on parameters such as leaf color, shape, and size. (Image: www.hs-osnabrueck.de)	PD
	Continental: Contadino [137]	250	RTK-GPS positioning system (3 cm accuracy), LIDAR sensors, 3D camera systems	Electrical motors	N.A	Seeding, weeding, spraying, sowing, fertilizing, monitoring. All plants in the field. (Image: www.continental.com)	PD
	Earth Rover: CLAWS	300	6 built-in cameras	Solar powered (12 panels) electrical motors	N.A	Concentrated light autonomous weeding and scouting (Image: www.earthrover.farm)	PD

TABLE 4.2 (Continued)
Summary of the Identified Field Robots in the European Market and Their Status of Development (SoD). CA: Commercially Available, UFT: Under Field Test, and PD: Prototype Development

Robot Image	Robot Name	Weight (kg)	Sensors	Motor Type/ Power (kW/hp)	Lifting Capacity	Function and Application Area	SoD
	Escarda Technologies: Escarda	N.A	Laser system	Solar-powered electrical motors	N.A	Sustainable laser-based weeding, organic crop production of tomatoes, sugar beet (Image: www.escarda.tech)	PD
	Fraunhofer IPA: CURT [138]	N.A	RTK-GPS, 3D LIDAR system	Electrical motors	N.A	2-wheel drive, CURT (crops under regular treatment) can perform chemical-free and sustainable weed regulation by mechanical methods, ridge crops, such as potatoes, onions or carrots (Image: www.iese.fraunhofer.de)	PD
	Gentle Robotics: E-Terry	N.A	N.A	N.A	N.A	Automated crop monitoring, soil analysis, weed control, individualized fertilization, automated harvesting (Image: www.e-terry.de)	PD

(Continued)

TABLE 4.2 (Continued)

Summary of the Identified Field Robots in the European Market and Their Status of Development (SoD). CA: Commercially Available, UFT: Under Field Test, and PD: Prototype Development

Robot Image	Robot Name	Weight (kg)	Sensors	Motor Type/ Power (kW/hp)	Lifting Capacity	Function and Application Area	SoD
	Innok Robotics: HEROS [139]	70–140	2D and 3D scanners, RTK-GNSS (0.5–2 cm accuracy), camera/vision and radar technologies	Electrical motors 0.8 (2WD) and 1.6 (4WD) (kW), 0.4–0.96 (kWh)	70–400	2- and 4-wheel drive, multifunction robot (spray robot, research robot, camera robot), mapping, surface processing. (Image: www.innok-robotics.de)	PD
	John Deere: Autonomous Field Sprayer [140]	2700	StarFire 6000 Receiver implementing an improved antenna, the latest in GNSS signal processing technology, and a differential correction signal	Combustion Engine	min. 560	Spraying with a 560-liter tank for arable crops, offers a high ground clearance of 1.9 m, which allows it to work in arable crops and open field vegetables. (Image: www.deere.de)	PD
	Strubes: PhenoBob [141]	1250	Combined colour and near-infrared cameras, RTK GPS	N.A	N.A	Monitoring and scouting, measurement of field emergence and early plant development, sugar beet. (Image: www.strube.net)	PD

TABLE 4.2 (Continued)
Summary of the Identified Field Robots in the European Market and Their Status of Development (SoD). CA: Commercially Available, UFT: Under Field Test, and PD: Prototype Development

Robot Image	Robot Name	Weight (kg)	Sensors	Motor Type/ Power (kW/hp)	Lifting Capacity	Function and Application Area	SoD
	Strubes: BlueBob [141]	N.A	6 camera systems	Electrical motors	N.A	2-wheel drive, hoeing and weeding, sugar beet. Strubes BlueBob was developed in cooperation with Naïo Technologies Fraunhofer. (Image: www.strube.net)	PD
	Small robot company: Harry [133]	N.A	N.A	Electrical motors	N.A	4-wheel drive, precision drilling, planting, arable crops, winter wheat (Image: www. smallrobotcompany.com)	PD
	Feldschwarm Technologies: Feldschwarm [142, 143]	5000–8000	GPS	Electrical motors	N.A	4-wheel drive, tillage and sowing, modular components with standardized interfaces (traction modules, storage of fertilizer and seeds, automation components) (Image: www.feldschwarm.de)	PD

FIGURE 4.16 A close view of the Schöneiche potato field located in the district Dahme-Spreewald in Brandenburg, Germany, showing (a) planting, (b) vegetative growth, (c) tuber maturation stage, and (d) harvesting.

TABLE 4.3
Summary of the Field Robotic Systems That Meet the Requirements for Potato Cultivation

Robot Name	Track Width (m)	Ground Clearance (m)	Ref
Robotti 150D	1.66 to 3.65	0.9*	[151]
Robotti LR	1.80 to 3.65	0.9*	[122, 152]
AgBot 5.115T2	1.5 to 3.2	0.42	[127]
Exobotic/LAND-A3 (WTD4)	1.5	N.A	[153]
Farming GT	1.35 to 2.25	0.5*	[109]
Orio	1.80 to 2.25	0.6*	[132]
Sitia: Trektor	1.08 to 2.44	Adjustable **	[124]
BoniRob	0.75 to 3	0.4 to 0.8	[136]
Strubes: BlueBob	1.8 to 2.25	0.6*	[141]

*Personal communication with the manufacturer. **No exact dimensions are available.

on the FD20 robot or by means of a tool carrier robot such as AgBot, Orio, Robotti LR, or Robotti 150D. However, due to its crawler tracks with a minimum width of 0.3 m [146], the AgBot can potentially injure the ridge structure. Therefore, the applicability of the AgBot needs to be tested in this regard. After weeding, preserving the ridge structure is crucial to prevent yield loss due to the occurrence of green potatoes [147]. Thus, re-ridging after mechanical weed control is very important and should be performed by the robot as well. This can be realized by using AgBot, Orio, or Robotti LR/150D and an equivalent tool carrier. Since potato cultivation is susceptible to erosion [148], it would be useful to combine the re-ridging with, for example, the formation of micro-dams. Table 4.3 summarizes the robot systems that were assessed as usable based on the cultivation requirements. For a field robot that can drive safely along ridges, a precise steering system is required that autonomously adapts its path to the ridge structures as they change due to erosion or plant development. An important optional function with regard to plant development would be the monitoring of plant health in conjunction with adapted pest or disease control and fertilization according to the precision agriculture concept. The automation of this task means a strong reduction in workload for the farmer as well as a possible reduction of fertilizer or pesticides in the range of 25% [149]. This can be realized by integrated camera systems and corresponding data processing functions in combination with corresponding sprayers. It may be considered to suspend the application of a field robot during the period of plant closure especially formed by potato varieties reaching tall plant heights. During this time, plant damage may not be completely avoidable using a field robot; therefore, drones can be used to take over monitoring or local spraying tasks here [150]. In summary, Robotti LR and Robotti 150D fulfill the determined minimum requirements for an application for potato cultivation. In addition, both field robots meet optional requirements with regard to functions of autonomous driving, spraying, and mechanical weed control if combined with the corresponding carrier tools.

4.7 CURRENT CHALLENGES AND LIMITATIONS OF AGRICULTURAL ROBOTICS

The reviewed literature on the use of robotic technology in agriculture reveals that the majority of the published works consider limitations in sensing and perception as the main challenges to developing an effective solution. This is due to the different shapes, sizes, colors, the texture of target objects, the position and orientation of target objects, and significant variations in environmental conditions such as light intensity, rain, dust, and dirt. It is extremely challenging for a robot to adapt to changes when presented with a new scenario in the field, and to be capable of switching between tasks for performing site-specific operations on individual plants. To address this, agricultural robots must have an active perception system that allows their sensing to adapt to unexpected conditions and effectively identify the location, orientation, and material properties of objects in complex environments where objects are only partially visible. This means that robots need to learn from previous experiences and build a world model that they use to reason about the environment and guide their active perception and planning and control of their arms and wheels.

On the topic of robotic manipulators, the main scientific and engineering challenges are to improve perception capabilities, control systems, gripping, and manipulation. Extensive research works are focused on sensor-driven applications, adapted for specific use cases such as harvesting and weeding. Although camera-based applications are considered a proven solution in the industry and are successfully used for object detection, localization, and even quality measures, the technology is not yet mature to be commercially employed in agriculture. The open question is to what extent can these solutions be applied to real field environments? Modern image processing, such as structure from motion [154, 155], allows 3D modeling of the observed scene, such as fruits in bushes, where two subsequent camera images are used to generate a 3D pointcloud. The main challenge involved here is to analyze this 3D scene towards the optimal viewing angle of the camera, or more precisely to analyze whether the robot arm and end-effector position are adequate, with enough free space, to reach for the fruit. A possible solution is to mount a camera on a 6- or 7-DoF redundant robot arm to generate the necessary series of images for closed-loop control. The resulting 3D models are then analyzed to define the ideal path for the end-effector of the robot. The target is an orthogonal picking direction for the addressed fruit. Additionally, the colors of the fruits can be analyzed for maturity level for selective fruit picking. This is crucial, especially in situations such as blueberry harvesting, where different grades of fruit maturity exist at the same time. This quality-oriented picking will balance the disadvantage of slower mechanical harvest because the harvest season can take the time up to several weeks. Published works demonstrate applications of different open-source software development tools for computer vision, machine learning, neuronal networks, AI, and ROS, which are interfaced with each other by means of MATLAB Simulink® and additional hardware support packages. The result allows establishing an embedded setup such as Jetson boards for a distributed image processing task without interfering via the main computer or ROS master.

Although soft robotics can be disruptive in how robot-human and robot-environment interactions are understood, this technology still has some challenges ahead with respect to soft manipulators and grippers. This is mainly due to the complexity of the unstructured agricultural environment, the intrinsic challenge posed by soft materials, and the need to demonstrate the economic viability of robotic harvesting in the sector [89]. Results of a recent study on the evaluation of a robust soft robotic gripper for apple harvesting in a natural orchard have shown a detachment, damage, and harvesting rate of 75.6%, 4.55%, and 70.77%, respectively [156]. Some of the main barriers that soft robotics, and more particularly soft grippers, face against their possible application in agriculture can be categorized as the design process [101], standardizing manufacturing methods [89], and features quantification (i.e., repeatability, reliability, and controllability). The design process is a major challenge in soft technology. Although a broad range of generalist soft grippers can be found in the current state of the art, they are primarily aimed at achieving new advancements in soft manipulators, rather than developing grippers that address specific application-related issues. When it comes to robotic crop harvesting, grippers that possess modular construction and are easy to repair and capable of handling various food crops are more desirable. Additionally, there is a need to examine the mathematical model that represents material behavior in FEM software, which presents another gap that requires attention. Standardizing manufacturing processes is a critical aspect to be addressed, as it would ensure that the designed soft actuators are suitable for production, thereby facilitating their integration into the robotic market. In this context, Vidwath et al. (2021) [89] have described the design and fabrication of a few soft robotic grippers using three types of liquid silicone rubber materials with different properties by fabricating one-fingered, three-fingered, and four-fingered soft pneumatic actuators with the potential to be used for harvesting in agricultural fields. Repeatability studies should focus on mitigating common issues that arise in soft actuators, such as delamination or interstitial bubbles, which may result from faulty manufacturing. Various solutions have been proposed to address these problems, including the use of vacuum chambers [157–160], which have shown positive results. However, it is difficult to find a method where variables such as pressure or time are controlled as a function of volume to guarantee the process's repeatability, which would ultimately depend on the material used. Additionally, most manufacturing processes are highly manual, which can compromise repeatability. Nevertheless, the future could see 3D printing of soft materials and lost wax manufacturing as promising options since they offer greater potential for achieving repeatability during the manufacturing process. For robotic crop harvesting, features such as modularity, ease of repair, and the ability to handle food and multiple crops are desired. On the control systems, the study of new control algorithms that take into account the stiffness of the object to be manipulated is essential for the implementation of soft technology in robotic crop harvesting.

Economic analysis plays a vital role in incentivizing research and development in any area, including Agriculture 4.0. In this field, economic studies can provide valuable insights into the most practical and profitable ways to harvest various crops. However, currently, there is a lack of economic research in this area. A 2019 study [107] pointed out that only 18 investigations have been carried out on estimating the profitability of crop automation. This has hindered the growth of automation technologies, including soft robotics, in the agricultural sector. Despite the lack of research, it is evident that labor cost at harvest time represents 30% of the total cost in certain crops, such as tomatoes and peppers [161]. Therefore, the use of soft grippers for mechanical harvesting may be a cost-effective alternative to manual harvesting [162]. To address the challenge of relatively slower actuation speeds, some studies are exploring the use of pneumatic channels, also known as pneumatic networks [163] or low-pressure actuators [164]. Additionally, hybrid gripper technology [165], which combines the advantages of soft and rigid robotics, may provide a solution by offering a soft grip and structural strength capable of withstanding external agents or objects present in unstructured environments.

For field robots, the main challenges toward wide adoption involve the economic feasibility in terms of capital investments, operating costs, and the need for personnel training [6, 166]. These robots often require substantial investment, high maintenance and repair costs, and specialized expertise and spare parts that may not be readily available in rural areas. Currently, most farms have not been optimized for field robot operations and lack the necessary infrastructure, such as compatible sensor networks, power supply systems, and communication infrastructure. On the other hand, retrofitting existing infrastructure to accommodate mobile robots can be expensive and time-consuming, further impacting the economic feasibility. From a technical perspective, field robots need to navigate through uneven surfaces, identify and manipulate crops with precision, and adapt to changing environmental conditions. Their autonomous navigation is also expected to demonstrate collision avoidance capabilities that can provide safe operation in actual field conditions. Achieving this level of adaptability poses significant technical challenges. Although some mobile robots are capable of operating in controlled environments, scaling their capabilities to handle the complexity of real-world agricultural settings remains a considerable hurdle. To overcome this, engineers are incorporating different sensing solutions, including RTK-GPS, inertial measurement unit (IMU), depth cameras, multi-spectral cameras, multi-channel distance detection sensors, and LiDARs to provide field robots with full autonomy [167, 168]. However, the development of a robust solution that can operate inside unstructured agricultural fields still proposes

serious challenges due to the extreme variations in high-density plants and disturbances of the outdoor environment. Data fusion and multiple perception solutions
are usually employed to assist the existing GPS-based navigation and to improve
the reliability and flexibility of the operation [169, 170]. For example, a robot may
be required to perform mowing between the plant rows with an accuracy of 5–10
cm from the side and an ideal speed of 5–8 km/h. Development of such a system
requires proof-of-concept via extensive validation tests that can be accelerated with
the digital representation of the sensors, dynamic models of the robot, and virtual
replicas of the orchard. More recently, machine learning (ML) and deep learning
(DL) techniques are being deployed as the control algorithms for autonomous navigation and collision avoidance of field robots to detect relationships, analyze patterns, and make predictions.

4.8 FUTURE SCENARIOS

With the introduction of the fifth-generation mobile network (5G), agriculture is
redefining some of the concepts of robotic technology. One of the trending topics in
this context is the deployment of distributed automation systems such as collaborative robots and swarms of small-scale unmanned machinery that can autonomously
execute various site-specific operations such as weeding and spraying via IoT-based
cloud computing services. An example includes the work of Zhang et al. (2021)
[171], in which a concept of vision-based navigation for agricultural IoT and a binocular vision navigation algorithm for agricultural robots is presented. Furthermore,
the robot gathers information on plant growth in real time, which can be sent to a
cloud computing server to be used in predicting yield and assessing crop health.
Although similar solutions have been already implemented as pilot plant projects or
on a commercial scale using different connectivity approaches such as 3G/4G, WiFi,
or LoRaWAN, the connection stability, limited availability, bandwidth, low latency,
and data security between nodes are always a concern [172].

Figure 4.17 shows a conceptual representation of intelligent connectivity in
action, applying 5G for machine-to-machine and machine-to-cloud data transfer via
an AI-hosted smart edge platform without human intervention, which can enhance
the decision-making and efficiency of agricultural robots. A key component to
enable this digital transformation is non-delayed data transfer between field/plant
sensors (i.e., microclimate, soil, leaf wetness, and fruit maturity), decision-making
models, and robots that respond to these decisions. With the high-speed data transfer
of up to 10 GB/s in the 5G network, it is possible to solve the problems existing in
the original wireless transmission [173] and provide a reliable and secure communication infrastructure with low latency capabilities for the realization of automated
farms by means of linking a large number of robots and data acquisition devices
[174–177]. Although 5G requires infrastructure and might take more time to cover
all remote areas, it will contribute significantly to reducing workforce requirements
with automation.

Recently, a case study has been conducted by KPN IoT Solutions (KPN, Rotterdam,
the Netherlands) and Wageningen University and Research (WUR) to develop an
IoT proof of concept that combines the application of 5G connected robots, artificial

intelligence, and cloud-based processing to address weed elimination in large-scale outdoor crop management [178]. It has been highlighted that AI algorithms will enable agricultural robots to make informed decisions on precise planting, irrigation, and harvesting schedules, optimizing resource usage and maximizing yields. In this study, image data are captured by the cameras and are sent to the KPN cloud-based edge computing server via the 5G connectivity, allowing different plant types to be identified by deep learning algorithms. The detected plant types are then sent as feedback to the robot to apply weed spray. The project reports achieving an accuracy of 95%, with the potential to cover 0.33 hectares per hour, which is significantly higher than the speed of manual operations. Similarly, the foodChain project, launched in 2022 at the Leibniz-Institut für Agrartechnik und Bioökonomie, addresses the challenge of further digitalization and automation by establishing a 5G campus network for real-time analysis of measurement data from the field in combination with real-time control of an autonomous mobile robot. This robot is being tested with AI controllers and path planning algorithms (i.e., A-star and Dijkstra's [179, 180]) in order to consider factors such as obstacle avoidance, distance, and time to find the most efficient route.

Governments worldwide have recognized the importance of digitalization in agriculture and have implemented initiatives to support the development and adoption of agricultural robotics. The SPARC initiative [12], among others, highlights the EU's direction for digitalization, aiming to enhance productivity, sustainability, and efficiency in the agricultural sector by focusing on fostering collaboration among industry stakeholders, researchers, and farmers to develop and deploy innovative solutions. The initiatives also provide funding opportunities, technical assistance, and knowledge-sharing platforms to facilitate the development and commercialization of agricultural robotics. These initiatives help the farmers' community to realize and trust the potential and benefits of robots in agricultural automation.

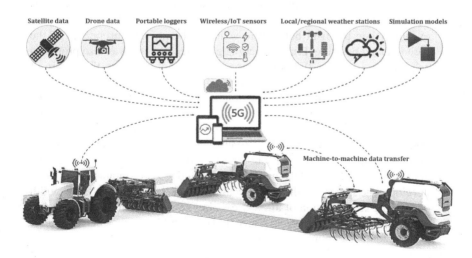

FIGURE 4.17 Connectivity of swarm of robots via 5G network. Image of the robots courtesy of www.iav.com.

4.9 CONCLUSION

This chapter provided an overview of the robotic manipulator, human-robot collaboration, soft-robotic technology, and field robots that are expected to revolutionize farming practices. The following conclusions can be summarized:

1. Despite the intensive studies on the use and control of robotic manipulators in agriculture, there are still no fully functional systems in the market that can replace the human workforce. Other than the safety regulations and costs, one of the most limiting factors for the widespread use of robots in agriculture lies in the existing sensing solutions that are not mature enough to generate precise information when facing different shapes and sizes of trees and plants, unstructured environments, varying outdoor weather and light conditions, and navigating on uneven terrains. Some studies have investigated the possibility of cooperation among multiple visual servoing methods, such as systems that consist of two or more independent fixed eye-to-hand and eye-in-hand cameras. Other studies have discussed the feasibility of human-robot collaboration, in which one or more human operators perform the sensing task of plant/fruit identification based on the live images, and the robots carry out the actuation tasks in the field.
2. Development of an affordable and efficient robotic system requires collaboration in areas of horticultural engineering, machine vision, sensing, robotics, control, intelligent systems, software architecture, system integration, and crop management. In addition, practicing other cultivation systems in the open field and closed field, such as a single row, might be necessary for overcoming the problems of plant/fruit visibility and accessibility. Moreover, the reviewed literature on fruit grasping in robotic manipulators shows that the gripping process in fruit harvesting is one of the most difficult tasks. The gripper or end-effector needs to have the dexterity not only to manipulate the fruit but also to adapt to the physical properties of the fruit to avoid bruising. Finger-tracking glove technology was also proposed as promising for the study of picking patterns and its use in the design of grippers adapted to the crop harvesting.
3. Based on the evaluation of the autonomous field robots available on the European market in potato cultivation, it was found that Robotti LR and Robotti 150 D can be considered two promising candidates for the automation of certain cultivation steps along the entire vegetation period. If the ground clearance of the BlueBob®, Orio, AgBot, and Farming GT could be adjusted to the identified 0.8 m, then they would also be suitable for all plant development stages. With some geometric adjustment in the track width of 5 cm or an expansion of already existing geometric variability on robot platforms such as Anatis from Carré, E-Tract, and Dino, they can also offer great potential for application in potato cultivation.
4. Some of the main limitations toward adaptation and commercialization of agricultural robotics are attributed to technological robustness and reliability, high costs of research and development, lack of IT infrastructure, and

lack of farmers' trust. Depending on the application, agricultural robotics has different stages of development that range from proof-of-concept to market-ready products. This review showed that the use of robots for applications such as weeding, which is constantly required during the entire season, is more attractive to farmers.

4.10   ACKNOWLEDGMENT

This study was conducted as part of the foodChain project funded by the Federal Ministry for Digital and Transport (BMDV). The authors would like to acknowledge the European Agricultural Fund for Rural Development (ELER) for funding the SunBot project. The authors also express their sincere gratitude for the valuable support provided by Mostafa Shokrian Zeini and Maryam Behjati, who contributed to the success of this research project.

4.11   DISCLAIMER

Mention of commercial products, services, trade or brand names, organizations, or research studies in this publication does not imply endorsement by the authors, nor discrimination against similar products, services, trade or brand names, organizations, or research studies not mentioned.

REFERENCES

[1]   U. Garlando, L. Bar-On, A. Avni, Y. Shacham-Diamand, and D. Demarchi, "Plants and Environmental Sensors for Smart Agriculture, an Overview," in *2020 IEEE SENSORS*, Rotterdam, Netherlands, 2020, pp. 1–4. https://doi.org/10.1109/SENSORS47125.2020.9278748

[2]   F. K. Shaikh, S. Karim, S. Zeadally, and J. Nebhen, "Recent Trends in Internet-of-Things-Enabled Sensor Technologies for Smart Agriculture," *IEEE Internet Things J.*, vol. 9, no. 23, pp. 23583–23598, 2022.

[3]   T. Wang, B. Chen, Z. Zhang, H. Li, and M. Zhang, "Applications of Machine Vision in Agricultural Robot Navigation: A Review," *Comput. Electron. Agric.*, vol. 198, p. 107085, 2022.

[4]   T. Ayoub Shaikh, T. Rasool, and F. Rasheed Lone, "Towards Leveraging the Role of Machine Learning and Artificial Intelligence in Precision Agriculture and Smart Farming," *Comput. Electron. Agric.*, vol. 198, p. 107119, 2022.

[5]   E. J. van Henten, A. Tabb, J. Billingsley, M. Popovic, M. Deng, and J. Reid, "Agricultural Robotics and Automation [TC Spotlight]," *IEEE Robot. Autom. Mag.*, vol. 29, no. 4, pp. 145–147, 2022.

[6]   R. Sparrow and M. Howard, "Robots in Agriculture: Prospects, Impacts, Ethics, and Policy," *Precis. Agric.*, vol. 22, no. 3, pp. 818–833, 2021.

[7]   J. Sung, "The Fourth Industrial Revolution and Precision Agriculture," *Autom. Agric. Secur. Food Supplies Futur. Gener.*, vol. 1, 2018.

[8]   D. S. Paraforos and H. W. Griepentrog, "Digital Farming and Field Robotics: Internet of Things, Cloud Computing, and Big Data," in *Fundamentals of Agricultural and Field Robotics*, M. Karkee and Q. Zhang, Eds. Cham: Springer International Publishing, 2021, pp. 365–385.

[9]   V. Sharma, A. K. Tripathi, and H. Mittal, "Technological Revolutions in Smart Farming: Current Trends, Challenges & Future Directions," *Comput. Electron. Agric.*, vol. 201, p. 107217, 2022.

[10]   D. Albiero, A. Pontin Garcia, C. Kiyoshi Umezu, and R. Leme de Paulo, "Swarm Robots in Mechanized Agricultural Operations: A Review about Challenges for Research," *Comput. Electron. Agric.*, vol. 193, p. 106608, 2022.

[11]   T. Martin et al., "Robots and Transformations of Work in Farm: A Systematic Review of the Literature and a Research Agenda," *Agron. Sustain. Dev.*, vol. 42, no. 4, p. 66, 2022.

[12]   T. Turja and A. Oksanen, "Robot Acceptance at Work: A Multilevel Analysis Based on 27 EU Countries," *Int. J. Soc. Robot.*, vol. 11, no. 4, pp. 679–689, 2019.

[13]   R. R. Shamshiri et al., "Research and Development in Agricultural Robotics: A Perspective of Digital Farming," *Int. J. Agric. Biol. Eng.*, vol. 11, no. 4, 2018.

[14]   G. Kootstra, A. Bender, T. Perez, and E. J. van Henten, "Robotics in Agriculture," in *Encyclopedia of Robotics*, M. H. Ang, O. Khatib, and B. Siciliano, Eds. Berlin, Heidelberg: Springer Berlin Heidelberg, 2020, pp. 1–19.

[15]   M. Bergerman, J. Billingsley, J. Reid, and E. van Henten, "Robotics in Agriculture and Forestry," in *Springer Handbook of Robotics*, B. Siciliano and O. Khatib, Eds. Cham: Springer International Publishing, 2016, pp. 1463–1492.

[16]   L. F. P. Oliveira, A. P. Moreira, and M. F. Silva, "Advances in Agriculture Robotics: A State-of-the-Art Review and Challenges Ahead," *Robotics*, vol. 10, no. 2, 2021.

[17]   L. N. K. Duong et al., "A Review of Robotics and Autonomous Systems in the Food Industry: From the Supply Chains Perspective," *Trends Food Sci. Technol.*, vol. 106, pp. 355–364, 2020.

[18]   A. Atefi, Y. Ge, S. Pitla, and J. Schnable, "Robotic Technologies for High-Throughput Plant Phenotyping: Contemporary Reviews and Future Perspectives," *Front. Plant Sci.*, vol. 12, 2021.

[19]   R. Xu and C. Li, "A Review of High-Throughput Field Phenotyping Systems: Focusing on Ground Robots," *Plant Phenomics*, vol. 2022, 2022.

[20]   L. Yao, R. van de Zedde, and G. Kowalchuk, "Recent Developments and Potential of Robotics in Plant Eco-Phenotyping," *Emerg. Top. Life Sci.*, vol. 5, no. 2, pp. 289–300, 2021.

[21]   L. Emmi and P. Gonzalez-de-Santos, "Mobile Robotics in Arable Lands: Current State and Future Trends," in *2017 European Conference on Mobile Robots (ECMR)*, Paris, France, 2017, pp. 1–6. https://doi.org/10.1109/ECMR.2017.8098694

[22]   G. Ren, T. Lin, Y. Ying, G. Chowdhary, and K. C. Ting, "Agricultural Robotics Research Applicable to Poultry Production: A Review," *Comput. Electron. Agric.*, vol. 169, p. 105216, 2020.

[23]   R. Barth, J. Hemming, and E. J. van Henten, "Design of an Eye-in-Hand Sensing and Servo Control Framework for Harvesting Robotics in Dense Vegetation," *Biosyst. Eng.*, vol. 146, pp. 71–84, 2016.

[24]   Q. Zhang, M. Karkee, and A. Tabb, "The Use of Agricultural Robots in Orchard Management," in *Robotics and Automation for Improving Agriculture*. Sawston, Cambridge: Burleigh Dodds Science Publishing, 2019, pp. 187–214.

[25]   L. F. P. Oliveira, A. P. Moreira, and M. F. Silva, "Advances in Forest Robotics: A State-of-the-Art Survey," *Robotics*, vol. 10, no. 2, 2021.

[26]   Y. Lu and S. Young, "A Survey of Public Datasets for Computer Vision Tasks in Precision Agriculture," *Comput. Electron. Agric.*, vol. 178, p. 105760, 2020.

[27]   S. Fountas, I. Malounas, L. Athanasakos, I. Avgoustakis, and B. Espejo-Garcia, "AI-Assisted Vision for Agricultural Robots," *AgriEngineering*, vol. 4, no. 3. pp. 674–694, 2022.

[28] H. Tian, T. Wang, Y. Liu, X. Qiao, and Y. Li, "Computer Vision Technology in Agricultural Automation—A Review," *Inf. Process. Agric.*, vol. 7, no. 1, pp. 1–19, 2020.

[29] S. A. Magalhães, A. P. Moreira, F. N. dos Santos, and J. Dias, "Active Perception Fruit Harvesting Robots—A Systematic Review," *J. Intell. Robot. Syst.*, vol. 105, no. 1, p. 14, 2022.

[30] L. C. Santos, F. N. Santos, E. J. S. Pires, A. Valente, P. Costa, and S. Magalhães, "Path Planning for Ground Robots in Agriculture: A Short Review," in *2020 IEEE International Conference on Autonomous Robot Systems and Competitions (ICARSC)*, Ponta Delgada, Portugal, 2020, pp. 61–66. https://doi.org/10.1109/ICARSC49921.2020.9096177

[31] J. F. Elfferich, D. Dodou, and C. D. Santina, "Soft Robotic Grippers for Crop Handling or Harvesting: A Review," *IEEE Access*, vol. 10, pp. 75428–75443, 2022.

[32] B. Zhang, Y. Xie, J. Zhou, K. Wang, and Z. Zhang, "State-of-the-Art Robotic Grippers, Grasping and Control Strategies, as Well as Their Applications in Agricultural Robots: A Review," *Comput. Electron. Agric.*, vol. 177, p. 105694, 2020.

[33] T. Yoshida, T. Kawahara, and T. Fukao, "Fruit Recognition Method for a Harvesting Robot with RGB-D Cameras," *ROBOMECH J.*, vol. 9, no. 1, p. 15, 2022.

[34] K. He, G. Gkioxari, P. Dollár, and R. Girshick, "Mask R-CNN," *IEEE Trans. Pattern Anal. Mach. Intell.*, vol. 42, no. 2, pp. 386–397, 2020.

[35] F. Tokuda, S. Arai, and K. Kosuge, "Convolutional Neural Network-Based Visual Servoing for Eye-to-Hand Manipulator," *IEEE Access*, vol. 9, pp. 91820–91835, 2021.

[36] D. M. Bulanon, C. Burr, M. DeVlieg, T. Braddock, and B. Allen, "Development of a Visual Servo System for Robotic Fruit Harvesting," *AgriEngineering*, vol. 3, no. 4, pp. 840–852, 2021.

[37] X. Ling, Y. Zhao, L. Gong, C. Liu, and T. Wang, "Dual-Arm Cooperation and Implementing for Robotic Harvesting Tomato Using Binocular Vision," *Rob. Auton. Syst.*, vol. 114, pp. 134–143, 2019.

[38] S. S. Mehta, W. MacKunis, and T. F. Burks, "Robust Visual Servo Control in the Presence of Fruit Motion for Robotic Citrus Harvesting," *Comput. Electron. Agric.*, vol. 123, pp. 362–375, 2016.

[39] C. W. Bac, J. Hemming, B. A. J. van Tuijl, R. Barth, E. Wais, and E. J. van Henten, "Performance Evaluation of a Harvesting Robot for Sweet Pepper," *J. F. Robot.*, vol. 34, no. 6, pp. 1123–1139, 2017.

[40] R. Fernández, H. Montes, J. Surdilovic, D. Surdilovic, P. Gonzalez-De-Santos, and M. Armada, "Automatic Detection of Field-Grown Cucumbers for Robotic Harvesting," *IEEE Access*, vol. 6, pp. 35512–35527, 2018.

[41] S. Mao, Y. Li, Y. Ma, B. Zhang, J. Zhou, and Kai Wang, "Automatic Cucumber Recognition Algorithm for Harvesting Robots in the Natural Environment Using Deep Learning and Multi-Feature Fusion," *Comput. Electron. Agric.*, vol. 170, p. 105254, 2020.

[42] Y. Zhao, L. Gong, Y. Huang, and C. Liu, "A Review of Key Techniques of Vision-Based Control for Harvesting Robot," *Comput. Electron. Agric.*, vol. 127, no. Supplement C, pp. 311–323, 2016.

[43] Y. Li, Q. Feng, T. Li, F. Xie, C. Liu, and Z. Xiong, "Advance of Target Visual Information Acquisition Technology for Fresh Fruit Robotic Harvesting: A Review," *Agronomy*, vol. 12, no. 6, 2022.

[44] A. AlBeladi, E. Ripperger, S. Hutchinson, and G. Krishnan, "Hybrid Eye-in-Hand/ Eye-to-Hand Image Based Visual Servoing for Soft Continuum Arms," *IEEE Robot. Autom. Lett.*, vol. 7, no. 4, pp. 11298–11305, 2022.

[45] P. Zapotezny-Anderson and C. Lehnert, "Towards Active Robotic Vision in Agriculture: A Deep Learning Approach to Visual Servoing in Occluded and Unstructured Protected Cropping Environments," *IFAC-PapersOnLine*, vol. 52, no. 30, pp. 120–125, 2019.

[46] D. I. Patrício and R. Rieder, "Computer Vision and Artificial Intelligence in Precision Agriculture for Grain Crops: A Systematic Review," *Comput. Electron. Agric.*, vol. 153, pp. 69–81, 2018.

[47] T. T. Santos, L. L. de Souza, A. A. dos Santos, and S. Avila, "Grape Detection, Segmentation, and Tracking Using Deep Neural Networks and Three-Dimensional Association," *Comput. Electron. Agric.*, vol. 170, p. 105247, 2020.

[48] M. M. Hasan, J. P. Chopin, H. Laga, and S. J. Miklavcic, "Detection and Analysis of Wheat Spikes Using Convolutional Neural Networks," *Plant Methods*, vol. 14, no. 1, pp. 1–13, 2018.

[49] Y. Majeed et al., "Deep Learning Based Segmentation for Automated Training of Apple Trees on Trellis Wires," *Comput. Electron. Agric.*, vol. 170, p. 105277, 2020.

[50] Z. Luo, W. Yang, Y. Yuan, R. Gou, and X. Li, "Semantic Segmentation of Agricultural Images: A Survey," *Information Processing in Agriculture*, vol. 11, no. 2, 2024, pp. 172–186, ISSN 2214-3173, https://doi.org/10.1016/j.inpa.2023.02.001. https://www.sciencedirect.com/science/article/pii/S2214317723000112

[51] F. Lateef and Y. Ruichek, "Survey on Semantic Segmentation Using Deep Learning Techniques," *Neurocomputing*, vol. 338, pp. 321–348, 2019.

[52] N. Marlier, O. Brüls, and G. Louppe, "Simulation-Based Bayesian Inference for Robotic Grasping," *arXiv Prepr. arXiv2303.05873*, 2023.

[53] D. Su, H. Kong, Y. Qiao, and S. Sukkarieh, "Data Augmentation for Deep Learning Based Semantic Segmentation and Crop-Weed Classification in Agricultural Robotics," *Comput. Electron. Agric.*, vol. 190, p. 106418, 2021.

[54] E. Aguilar, B. Nagarajan, B. Remeseiro, and P. Radeva, "Bayesian Deep Learning for Semantic Segmentation of Food Images," *Comput. Electr. Eng.*, vol. 103, p. 108380, 2022.

[55] R. R. Shamshiri, I. A. Hameed, M. Karkee, and C. Weltzien, "Robotic Harvesting of Fruiting Vegetables: A Simulation Approach in V-REP, ROS and MATLAB," in *Automation in Agriculture-Securing Food Supplies for Future Generations*. London: InTech, 2018. https://doi.org/10.5772/intechopen.73861

[56] L. Fu, F. Gao, J. Wu, R. Li, M. Karkee, and Q. Zhang, "Application of Consumer RGB-D Cameras for Fruit Detection and Localization in Field: A Critical Review," *Comput. Electron. Agric.*, vol. 177, p. 105687, 2020.

[57] J. Gené-Mola et al., "Assessing the Performance of RGB-D Sensors for 3D Fruit Crop Canopy Characterization under Different Operating and Lighting Conditions," *Sensors*, vol. 20, no. 24, 2020.

[58] A. Beyaz, "Accuracy Detection of Intel® Realsense D455 Depth Camera for Agricultural Applications," XIII International Scientific Agricultural Symposium "Agrosym 2022", *Sarajevo, Bosnia And Herzegovina*, 6 - 09 October 2022, pp. 185 -194. Publication Type: Conference Paper / Full Text. Sarajevo: Bosnia And Herzegovina. https://agrosym.ues.rs.ba/article/showpdf/BOOK_OF_PROCEEDINGS_2022.pdf

[59] R. R. Shamshiri et al., "Simulation Software and Virtual Environments for Acceleration of Agricultural Robotics: Features Highlights and Performance Comparison," *Int. J. Agric. Biol. Eng.*, vol. 11, no. 4, 2018.

[60] D. Le Hanh, N. Van Luat, and L. N. Bich, "Combining 3D Matching and Image Moment Based Visual Servoing for Bin Picking Application," *Int. J. Interact. Des. Manuf.*, pp. 1–9, 2022.

[61] A. Zahid, M. S. Mahmud, L. He, P. Heinemann, D. Choi, and J. Schupp, "Technological Advancements Towards Developing a Robotic Pruner for Apple Trees: A Review," *Comput. Electron. Agric.*, vol. 189, p. 106383, 2021.

[62] R. Verbiest, K. Ruysen, T. Vanwalleghem, E. Demeester, and K. Kellens, "Automation and Robotics in the Cultivation of Pome Fruit: Where Do We Stand Today?," *J. F. Robot.*, vol. 38, no. 4, pp. 513–531, 2021.

[63]   N. Strisciuglio et al., "TrimBot2020: An Outdoor Robot for Automatic Gardening," *ISR 2018; 50th International Symposium on Robotics*, Munich, Germany, 2018, pp. 1–6. https://arxiv.org/abs/1804.01792

[64]   B. M. van Marrewijk et al., "Evaluation of a boxwood topiary trimming robot," *Biosyst. Eng.*, vol. 214, pp. 11–27, 2022.

[65]   L. Zhang et al., "A Quadratic Traversal Algorithm of Shortest Weeding Path Planning for Agricultural Mobile Robots in Cornfield," *J. Robot.*, vol. 2021, p. 6633139, 2021.

[66]   I. Vatavuk, G. Vasiljević, and Z. Kovačić, "Task Space Model Predictive Control for Vineyard Spraying with a Mobile Manipulator," *Agriculture*, vol. 12, no. 3. 2022.

[67]   R. Oberti et al., "Selective Spraying of Grapevines for Disease Control Using a Modular Agricultural Robot," *Biosyst. Eng.*, vol. 146, pp. 203–215, 2016.

[68]   T. Yoshida, Y. Onishi, T. Kawahara, and T. Fukao, "Automated Harvesting by a Dual-Arm Fruit Harvesting Robot," *ROBOMECH J.*, vol. 9, no. 1, p. 19, 2022.

[69]   C. Lehnert, C. McCool, I. Sa, and T. Perez, "Performance Improvements of a Sweet Pepper Harvesting Robot in Protected Cropping Environments," *J. F. Robot.*, vol. 37, no. 7, pp. 1197–1223, 2020.

[70]   K. Zhang, K. Lammers, P. Chu, Z. Li, and R. Lu, "System Design and Control of an Apple Harvesting Robot," *Mechatronics*, vol. 79, p. 102644, 2021.

[71]   H. Kang, H. Zhou, and C. Chen, "Visual Perception and Modeling for Autonomous Apple Harvesting," *IEEE Access*, vol. 8, pp. 62151–62163, 2020.

[72]   D. M. Bulanon, H. Okamoto, and S.-I. Hata, *Feedback Control of Manipulator Using Machine Vision for Robotic Apple Harvesting*. St. Joseph, MI: ASAE, 2005.

[73]   Z. De-An, L. Jidong, J. Wei, Z. Ying, and C. Yu, "Design and Control of an Apple Harvesting Robot," *Biosyst. Eng.*, vol. 110, no. 2, pp. 112–122, 2011.

[74]   Y. Onishi, T. Yoshida, H. Kurita, T. Fukao, H. Arihara, and A. Iwai, "An Automated Fruit Harvesting Robot by Using Deep Learning," *ROBOMECH J.*, vol. 6, no. 1, p. 13, 2019.

[75]   S. S. Mehta and T. F. Burks, "Vision-Based Control of Robotic Manipulator for Citrus Harvesting," *Comput. Electron. Agric.*, vol. 102, pp. 146–158, 2014.

[76]   D. SepúLveda, R. Fernández, E. Navas, M. Armada, and P. González-De-Santos, "Robotic Aubergine Harvesting Using Dual-Arm Manipulation," *IEEE Access*, vol. 8, pp. 121889–121904, 2020.

[77]   G. Adamides et al., "HRI Usability Evaluation of Interaction Modes for a Teleoperated Agricultural Robotic Sprayer," *Appl. Ergon.*, vol. 62, pp. 237–246, 2017.

[78]   J. P. Vasconez, G. A. Kantor, and F. A. Auat Cheein, "Human–Robot Interaction in Agriculture: A Survey and Current Challenges," *Biosyst. Eng.*, vol. 179, pp. 35–48, 2019.

[79]   G. Adamides et al., "Design and Development of a Semi-Autonomous Agricultural Vineyard Sprayer: Human–Robot Interaction Aspects," *J. F. Robot.*, vol. 34, no. 8, pp. 1407–1426, 2017.

[80]   C. Lytridis et al., "An Overview of Cooperative Robotics in Agriculture," *Agronomy*, vol. 11, no. 9, 2021.

[81]   M. Polic, M. Car, F. Petric, and M. Orsag, "Compliant Plant Exploration for Agricultural Procedures with a Collaborative Robot," *IEEE Robot. Autom. Lett.*, vol. 6, no. 2, pp. 2768–2774, 2021.

[82]   E. Tziolas et al., "Assessing the Economic Performance of Multipurpose Collaborative Robots toward Skillful and Sustainable Viticultural Practices," *Sustainability*, vol. 15, no. 4, 2023.

[83]   E. Tziolas et al., "Comparative Assessment of Environmental/Energy Performance under Conventional Labor and Collaborative Robot Scenarios in Greek Viticulture," *Sustainability*, vol. 15, no. 3, 2023.

[84]   R. Berenstein and Y. Edan, "Human-Robot Collaborative Site-Specific Sprayer," *J. F. Robot.*, vol. 34, no. 8, pp. 1519–1530, 2017.

[85]   L. Rosalia et al., "A Soft Robotic Sleeve for Compression Therapy of the Lower Limb," in *2021 43rd Annual International Conference of the IEEE Engineering in Medicine & Biology Society (EMBC)*, Mexico, 2021, pp. 1280–1283. https://doi.org/10.1109/ EMBC46164.2021.9630924

[86]   P. Polygerinos, Z. Wang, K. C. Galloway, R. J. Wood, and C. J. Walsh, "Soft Robotic Glove for Combined Assistance and at-Home Rehabilitation," *Rob. Auton. Syst.*, vol. 73, pp. 135–143, 2015.

[87]   C. Firth, K. Dunn, M. H. Haeusler, and Y. Sun, "Anthropomorphic Soft Robotic End-Effector for Use with Collaborative Robots in the Construction Industry," *Autom. Constr.*, vol. 138, p. 104218, 2022.

[88]   R. Morales, F. J. Badesa, N. Garcia-Aracil, J. M. Sabater, and L. Zollo, "Soft Robotic Manipulation of Onions and Artichokes in the Food Industry," *Adv. Mech. Eng.*, vol. 6, p. 345291, 2014.

[89]   S. M. G. Vidwath, P. Rohith, R. Dikshithaa, N. Nrusimha Suraj, R. G. Chittawadigi, and M. Sambandham, "Soft Robotic Gripper for Agricultural Harvesting," in *Machines, Mechanism and Robotics*, R. Kumar, V. S. Chauhan, M. Talha, and H. Pathak, Eds. Singapore: Springer, 2022, pp. 1347–1353. https://doi.org/10.1007/978-981-16-0550-5_128

[90]   E. Navas, R. Fernández, D. Sepúlveda, M. Armada, and P. Gonzalez-de-Santos, "Soft Grippers for Automatic Crop Harvesting: A Review," *Sensors*, vol. 21, no. 8, 2021.

[91]   E. Navas, R. Fernández, M. Armada, and P. Gonzalez-de-Santos, "Diaphragm-Type Pneumatic-Driven Soft Grippers for Precision Harvesting," *Agronomy*, vol. 11, no. 9, 2021.

[92]   G. Chowdhary, M. Gazzola, G. Krishnan, C. Soman, and S. Lovell, "Soft Robotics as an Enabling Technology for Agroforestry Practice and Research," *Sustainability*, vol. 11, no. 23, 2019.

[93]   T. George Thuruthel, Y. Ansari, E. Falotico, and C. Laschi, "Control Strategies for Soft Robotic Manipulators: A Survey," *Soft Robot.*, vol. 5, no. 2, pp. 149–163, 2018.

[94]   S. Neppalli et al., "OctArm—A Soft Robotic Manipulator," in *2007 IEEE/RSJ International Conference on Intelligent Robots and Systems*, San Diego, CA, 2007, p. 2569. https://doi.org/10.1109/IROS.2007.4399146

[95]   F. Renda, M. Giorelli, M. Calisti, M. Cianchetti, and C. Laschi, "Dynamic Model of a Multibending Soft Robot Arm Driven by Cables," *IEEE Trans. Robot.*, vol. 30, no. 5, pp. 1109–1122, 2014.

[96]   J. Zhou, S. Chen, and Z. Wang, "A Soft-Robotic Gripper with Enhanced Object Adaptation and Grasping Reliability," *IEEE Robot. Autom. Lett.*, vol. 2, no. 4, pp. 2287–2293, 2017.

[97]   Y. Hao et al., "Universal Soft Pneumatic Robotic Gripper with Variable Effective Length," in *2016 35th Chinese Control Conference (CCC)*, Chengdu, China, 2016, pp. 6109–6114. https://doi.org/10.1109/ChiCC.2016.7554316

[98]   K. C. Galloway et al., "Soft Robotic Grippers for Biological Sampling on Deep Reefs," *Soft Robot.*, vol. 3, no. 1, pp. 23–33, 2016.

[99]   A. Galley, G. K. Knopf, and M. Kashkoush, "Pneumatic Hyperelastic Actuators for Grasping Curved Organic Objects," *Actuators*, vol. 8, no. 4, p. 76, 2019.

[100] G. Carabin, I. Palomba, D. Matt, and R. Vidoni, "Experimental Evaluation and Comparison of Low-Cost Adaptive Mechatronic Grippers," in *Advances in Service and Industrial Robotics. RAAD 2017*. Mechanisms and Machine Science, vol. 49, C. Ferraresi and G. Quaglia, Eds. Cham: Springer, 2018, pp. 630–637. https://doi.org/10.1007/978-3-319-61276-8_66

[101] C. J. Hohimer, H. Wang, S. Bhusal, J. Miller, C. Mo, and M. Karkee, "Design and Field Evaluation of a Robotic Apple Harvesting System with a 3D-Printed Soft-Robotic End-Effector," *Trans. ASABE*, vol. 62, no. 2, pp. 405–414, 2019.

[102] K. Chen et al., "A Soft Gripper Design for Apple Harvesting with Force Feedback and Fruit Slip Detection," *Agriculture*, vol. 12, no. 11, 2022.

[103] A. L. Gunderman, J. A. Collins, A. L. Myers, R. T. Threlfall, and Y. Chen, "Tendon-Driven Soft Robotic Gripper for Blackberry Harvesting," *IEEE Robot. Autom. Lett.*, vol. 7, no. 2, pp. 2652–2659, 2022.

[104] U. Jeong, K. Kim, S.-H. Kim, H. Choi, B. D. Youn, and K.-J. Cho, "Reliability Analysis of a Tendon-Driven Actuation for Soft Robots," *Int. J. Rob. Res.*, vol. 40, no. 1, pp. 494–511, 2020.

[105] A. Beyaz, "Harvest Glove and LabView Based Mechanical Damage Determination on Apples," *Sci. Hortic. (Amsterdam)*, vol. 228, pp. 49–55, 2018.

[106] F. Rübcke von Veltheim and H. Heise, "German Farmers' Attitudes on Adopting Autonomous Field Robots: An Empirical Survey," *Agriculture*, vol. 11, no. 3, p. 216, 2021.

[107] J. Lowenberg-DeBoer, I. Y. Huang, V. Grigoriadis, and S. Blackmore, "Economics of Robots and Automation in Field Crop Production," *Precis. Agric.*, vol. 21, no. 2, pp. 278–299, 2020.

[108] G. B. Verne, "Adapting to a Robot: Adapting Gardening and the Garden to Fit a Robot Lawn Mower," in *Companion of the 2020 ACM/IEEE International Conference on Human-Robot Interaction (HRI '20)*. New York: Association for Computing Machinery, 2020, pp. 34–42. https://doi.org/10.1145/3371382.3380738

[109] R. Gerhards, D. Andújar Sanchez, P. Hamouz, G. G. Peteinatos, S. Christensen, and C. Fernandez-Quintanilla, "Advances in Site-Specific Weed Management in Agriculture—A Review," *Weed Res.*, vol. 62, no. 2, pp. 123–133, 2022.

[110] A. T. Meshram, A. V Vanalkar, K. B. Kalambe, and A. M. Badar, "Pesticide Spraying Robot for Precision Agriculture: A Categorical Literature Review and Future Trends," *J. F. Robot.*, vol. 39, no. 2, pp. 153–171, 2022.

[111] P. Karpyshev, V. Ilin, I. Kalinov, A. Petrovsky, and D. Tsetserukou, "Autonomous Mobile Robot for Apple Plant Disease Detection Based on CNN and Multi-Spectral Vision System," in *2021 IEEE/SICE International Symposium on System Integration (SII)*, Iwaki, Fukushima, Japan, 2021, pp. 157–162. https://doi.org/10.1109/IEEECONF49454.2021.9382649

[112] Y. Zhang, E. S. Staab, D. C. Slaughter, D. K. Giles, and D. Downey, "Automated Weed Control in Organic Row Crops Using Hyperspectral Species Identification and Thermal Micro-Dosing," *Crop Prot.*, vol. 41, pp. 96–105, 2012.

[113] D. Tiozzo Fasiolo, L. Scalera, E. Maset, and A. Gasparetto, "Recent Trends in Mobile Robotics for 3D Mapping in Agriculture," in *Advances in Service and Industrial Robotics. RAAD 2022. Mechanisms and Machine Science*, vol. 120, A. Müller and M. Brandstötter, Eds. Cham: Springer, 2022, pp. 428–435.

[114] D. Q. da Silva, F. N. Dos Santos, A. J. Sousa, and V. Filipe, "Visible and Thermal Image-Based Trunk Detection with Deep Learning for Forestry Mobile Robotics," *J. Imaging*, vol. 7, no. 9, p. 176, 2021.

[115] A. Milella, G. Reina, and M. Nielsen, "A Multi-Sensor Robotic Platform for Ground Mapping and Estimation Beyond the Visible Spectrum," *Precis. Agric.*, vol. 20, no. 2, pp. 423–444, 2019.

[116] Y. Yamasaki, M. Morie, and N. Noguchi, "Development of a High-Accuracy Autonomous Sensing System for a Field Scouting Robot," *Comput. Electron. Agric.*, vol. 193, p. 106630, 2022.

[117] P. Mishra, G. Polder, and N. Vilfan, "Close Range Spectral Imaging for Disease Detection in Plants Using Autonomous Platforms: A Review on Recent Studies," *Curr. Robot. Reports*, vol. 1, no. 2, pp. 43–48, 2020.

[118] P. Kurtser, O. Ringdahl, N. Rotstein, R. Berenstein, and Y. Edan, "In-Field Grape Cluster Size Assessment for Vine Yield Estimation Using a Mobile Robot and a Consumer Level RGB-D Camera," *IEEE Robot. Autom. Lett.*, vol. 5, no. 2, pp. 2031–2038, 2020.

[119] J. Massah, K. A. Vakilian, M. Shabanian, and S. M. Shariatmadari, "Design, Development, and Performance Evaluation of a Robot for Yield Estimation of Kiwifruit," *Comput. Electron. Agric.*, vol. 185, p. 106132, 2021.

[120] R. R. Shamshiri, C. Weltzien, and T. Schutte, "Multi-Sensor Data Fusion with Fuzzy Knowledge-based Controller for Collision Avoidance of a Mobile Robot," in *LAND. TECHNIK 2022: The Forum for Agricultural Engineering Innovations*, 1st ed., VDI Wissensforum GmbH, Ed. Düsseldorf: VDI Verlag, 2022, pp. 349–358.

[121] R. Shamshiri, C. Weltzien, I. Zytoon, and B Sakal, "Evaluation of Laser and Infrared Sensors with CANBUS Communication for Collision Avoidance of a Mobile Robot," in *Proceedings International Conference on Agricultural Engineering. AgEng-LAND. TECHNIK 2022.* Düsseldorf: VDI Verlag GmbH (0083-5560/978-3-18092406-9), 2022, pp. 121–130. https://www.vdi-nachrichten.com/shop/ageng-land-technik-2022/

[122] G. Luppi, "Autonomous Robot for Planting for Great Green Wall Initiative Feasibility Study," *Politecnico Milano*, 2022. https://www.politesi.polimi.it/retrieve/f1f6c879-dc62-4262-a966-ba994215d78a/Autonomous%20Robot%20for%20Planting%20for%20Great%20Green%20Wall%20Initiative%20Feasibility%20Study.pdf

[123] Q. Yang, X. Du, Z. Wang, Z. Meng, Z. Ma, and Q. Zhang, "A Review of Core Agricultural Robot Technologies for Crop Productions," *Comput. Electron. Agric.*, vol. 206, p. 107701, 2023. https://doi.org/10.1016/j.compag.2023.107701. https://www.sciencedirect.com/science/article/pii/S0168169923000893

[124] H. Nehme, C. Aubry, T. Solatges, X. Savatier, R. Rossi, and R. Boutteau, "LiDAR-based Structure Tracking for Agricultural Robots: Application to Autonomous Navigation in Vineyards," *J. Intell. Robot. Syst.*, vol. 103, no. 4, p. 61, 2021.

[125] I. Bručienė, S. Buragienė, and E. Šarauskis, "Weeding Effectiveness and Changes in Soil Physical Properties Using Inter-Row Hoeing and a Robot," *Agronomy*, vol. 12, no. 7, 2022.

[126] S. D. Martínez Castillo, "Interacting Imaginaries in the Making of an Alternative Future in Agriculture," *The Hague*, The Netherlands, 2021. https://thesis.eur.nl/pub/61247/MartA-nez_Castillo-_Sergio_David_MA_2020_21_AFES.pdf

[127] J. L. Lourenço, L. Conde Bento, A. P. Coimbra, and A. T. De Almeida, "Comparative Evaluation of Mobile Platforms for Non-Structured Environments and Performance Requirements Identification for Forest Clearing Applications," *Forests*, vol. 13, no. 11, 2022.

[128] M. Jasiński, J. Mączak, P. Szulim, and S. Radkowski, "Autonomous Agricultural Robot—Testing of the Vision System for Plants/Weed Classification," in *Automation 2018*. Advances in Intelligent Systems and Computing, vol. 743, R. Szewczyk, C. Zieliński, and M. Kaliczyńska, Eds. Cham: Springer, 2018, pp. 473–482. https://doi.org/10.1007/978-3-319-77179-3_44

[129] L. Lac, "Méthodes de vision par ordinateur et d'apprentissage profond pour la localisation, le suivi et l'analyse de structure de plantes: application au désherbage de précision," *Automatique*. Université de Bordeaux, 2022. Français. https://theses.hal.science/tel-03623086/

[130] C. Robert, T. Sotiropoulos, H. Waeselynck, J. Guiochet, and S. Vernhes, "The Virtual Lands of Oz: Testing an Agribot in Simulation," *Empir. Softw. Eng.*, vol. 25, no. 3, pp. 2025–2054, 2020.

[131] G. Gil, D. E. Casagrande, L. P. Cortés, and R. Verschae, "Why the Low Adoption of Robotics in the Farms? Challenges for the Establishment of Commercial Agricultural Robots," *Smart Agric. Technol.*, vol. 3, p. 100069, 2023.

[132] T. Herlitzius, M. Hengst, T. Bögel, S. Geißler, and S. Schwede, "Bodenbearbeitungstechnik," *Bodenbearbeitungstechnik. Jahrb. Agrartech.*, 2021. https://leopard.tu-braunschweig.de/servlets/MCRFileNodeServlet/dbbs_derivate_00049357/jahrbuchagrartechnik2021_bodenbearbeitung.pdf

[133] C. McLellan, "Smart Farming: How IoT, Robotics, and AI are Tackling One of the Biggest Problems of the Century," *TechRepublic*, 2018. https://utechnologies.co.za/wp-content/uploads/2019/01/IoT-Robotics-and-AI-Enabling-Farming.pdf

[134] V. Czymmek, L. O. Harders, F. J. Knoll, and S. Hussmann, "Accuracy Evaluation of a Weeding Robot in Organic Farming," in *2022 IEEE International Instrumentation and Measurement Technology Conference (I2MTC)*, Ottawa, ON, Canada, 2022, pp. 1–6. https://doi.org/10.1109/I2MTC48687.2022.9806571

[135] G. Cisek, "How Is AI Being Realized?: AI Determines Our Lives," in *The Triumph of Artificial Intelligence: How Artificial Intelligence Is Changing the Way We Live Together*, G. Cisek, Ed. Wiesbaden: Springer Fachmedien Wiesbaden, 2021, pp. 57–91.

[136] A. Ruckelshausen et al., "BoniRob–an Autonomous Field Robot Platform for Individual Plant Phenotyping," *Precis. Agric.*, vol. 9, no. 841, p. 1, 2009.

[137] C. Apachite, R. Lauxmann, R. Thiel, and A. Ratte-Front, "AI for Automated Driving," *ATZelectronics Worldw.*, vol. 16, no. 9, pp. 48–51, 2021.

[138] J. Osten, C. Weyers, K. Bregler, T. Emter, and J. Petereit, "Modular and Scalable Automation for Field Robots," *at - Automatisierungstechnik*, vol. 69, no. 4, pp. 307–315, 2021. https://doi.org/10.1515/auto-2020-0039

[139] R. K. Megalingam, A. H. Kota, V. K. T. Puchakayala, and A. S. Ganesh, "Analysis and Performance Evaluation of Innok Heros Robot," in *Inventive Systems and Control*. Lecture Notes in Networks and Systems, vol. 204, V. Suma, J. I. Z. Chen, Z. Baig, and H. Wang, Eds. Singapore: Springer, 2021, pp. 447–460. https://doi.org/10.1007/978-981-16-1395-1_33

[140] T. A. Burgers, J. D. Gaard, and B. J. Hyronimus, "Comparison of Three Commercial Automatic Boom Height Systems for Agricultural Sprayers," *Appl. Eng. Agric.*, vol. 37, no. 2, pp. 287–298, 2021.

[141] D. Herrmann, E.-M. Dillschneider, J.-U. Niemann, M. Tomforde, and J. K. Wegener, "Innovationen in der Pflanzenschutztechnik," *Jahrb. Agrartech. 2021*, vol. 33, 2022.

[142] S. Schwich, I. Stasewitsch, M. Fricke, and J. Schattenberg, "Übersicht zur Feld-Robotik in der Landtechnik," in *Jahrbuch Agrartechnik 2018*, L. Frerichs, Ed. Braunschweig, Germany: Institut für mobile Maschinen und Nutzfahrzeug, 2018.

[143] T. Herlitzius, H. Fichtl, A. Grosa, M. Henke, and M. Hengst, "Feldschwarm–Modular and Scalable Tillage Systems with Shared Autonomy," *LAND. Tech. AgEng*, pp. 409–420, 2019.

[144] J. Käthner, K. Koch, N. Höfner, V. Dworak, M. Shokrian Zeini, R. Shamshiri, W. Figurski, and C. Weltzien, "Review of Agricultural Field Robots and Their Applicability in Potato Cultivation," in *Resiliente Agri-Food-Systeme. Referate der 43. GIL-Jahrestagung. 43. GIL-Jahrestagung, Resiliente Agri-Food-Systeme: Herausforderungen und Lösungsansätze*, C. Hoffmann, A. Stein, A. Ruckelshausen, H. Müller, T. Steckel, and H. Floto, Eds. Bonn: Gesellschaft für Informatik (1617-5468/978-3-88579-724-1), 2023, pp. 137–148. https://gil-net.de/wp-content/uploads/2023/01/GIL_2023_Tagungsband_final_8.2.pdf

[145] Y. Tadesse, C. J. M. Almekinders, R. P. O. Schulte, and P. C. Struik, "Understanding Farmers' Potato Production Practices and Use of Improved Varieties in Chencha, Ethiopia," *J. Crop Improv.*, vol. 31, no. 5, pp. 673–688, 2017.

[146] S. Raikwar, J. Fehrmann, and T. Herlitzius, "Navigation and Control Development for a Four-Wheel-Steered Mobile Orchard Robot Using Model-Based Design," *Comput. Electron. Agric.*, vol. 202, p. 107410, 2022.

[147] F. Vučajnk, M. Vidrih, and R. Bernik, "Physical and Mechanical Properties of Soil for Ridge Formation, Ridge Geometry and Yield in New Planting and Ridge Formation Methods of Potato Production," *Irish J. Agric. Food Res.*, pp. 13–31, 2012.

[148] A. A. Ustroyev and E. A. Murzaev, "Influence of Crops of the Cover Crop When Forming the Ridge Surface of Landings of Potatoes to Dynamics of Parameters of the Soil State," *E3S Web of Conferences*, vol. 262, p. 1038, 2021.

[149] C. Kempenaar, T. Been, J. Booij, F. van Evert, J.-M. Michielsen, and C. Kocks, "Advances in Variable Rate Technology Application in Potato in the Netherlands," *Potato Res.*, vol. 60, no. 3, pp. 295–305, 2017.

[150] T. Schütte, V. Dworak, and C. Weltzien, "Deriving Precise Orchard Maps for Unmanned Ground Vehicles from UAV Images," in *Informatik in der Land-, Forst- und Ernährungswirtschaft. Fokus: Künstliche Intelligenz in der Agrar- und Ernährungswirtschaft. Referate der 42. GIL-Jahrestagung. 42. GIL-Jahrestagung, Künstliche Intelligenz der Agrar. Ernährungswirtschaft*, M. Gandorfer, C. Hoffmann, N. El Benni, M. Cockburn, T. Anken, and H. Floto, Eds. Bonn: Gesellschaft für Informatik (1617-5468/978-3-88579-711-1), 2022, pp. 271–276. https://gil-net.de/wp-content/uploads/2022/02/GIL-Tagungsband2022.pdf

[151] A. Calleja Huerta, M. Lamandé, O. Green, and L. Juhl Munkholm, "Effects of Load and Repeated Wheeling from Lightweight Autonomous Field Robots on Soil Structure," in *EGU General Assembly 2022*, Vienna, Austria, 23–27 May 2022, pp. EGU22-3581. https://doi.org/10.5194/egusphere-egu22-3581. https://meetingorganizer.copernicus.org/EGU22/EGU22-3581.html?pdf

[152] T. Herlitzius, M. Hengst, T. Bögel, S. Geißler, and S. Schwede, "Bodenbearbeitungstechnik," *Jahrbuch Agrartechnik*, 2021. https://leopard.tu-braunschweig.de/servlets/MCRFile NodeServlet/dbbs_derivate_00049357/jahrbuchagrartechnik2021_bodenbearbeitung.pdf

[153] Exobotic, "LAND-A3 (WTD4)" [Online]. Available: https://exobotic.com/products/L/58374093-d1cb-4127-a5ea-ccc7226a9385/ [Accessed: 12-Dec-2022].

[154] S. Jay, G. Rabatel, X. Hadoux, D. Moura, and N. Gorretta, "In-Field Crop Row Phenotyping from 3D Modeling Performed Using Structure from Motion," *Comput. Electron. Agric.*, vol. 110, pp. 70–77, 2015.

[155] J. Gené-Mola et al., "Fruit Detection and 3D Location Using Instance Segmentation Neural Networks and Structure-from-Motion Photogrammetry," *Comput. Electron. Agric.*, vol. 169, p. 105165, 2020.

[156] X. Wang, H. Kang, H. Zhou, W. Au, M. Y. Wang, and C. Chen, "Development and Evaluation of a Robust Soft Robotic Gripper for Apple Harvesting," *Comput. Electron. Agric.*, vol. 204, p. 107552, 2023.

[157] C. B. Teeple, T. N. Koutros, M. A. Graule, and R. J. Wood, "Multi-Segment Soft Robotic Fingers Enable Robust Precision Grasping," *Int. J. Rob. Res.*, vol. 39, no. 14, pp. 1647–1667, 2020.

[158] J. Friend and L. Yeo, "Fabrication of Microfluidic Devices Using Polydimethylsiloxane," *Biomicrofluidics*, vol. 4, no. 2, p. 26502, 2010.

[159] G. Rateni, M. Cianchetti, G. Ciuti, A. Menciassi, and C. Laschi, "Design and Development of a Soft Robotic Gripper for Manipulation in Minimally Invasive Surgery: A Proof of Concept," *Meccanica*, vol. 50, pp. 2855–2863, 2015.

[160] C. Linghu et al., "Universal SMP Gripper with Massive and Selective Capabilities for Multiscaled, Arbitrarily Shaped Objects," *Sci. Adv.*, vol. 6, no. 7, p. eaay5120, 2020.

[161] L. Droukas et al., "A Survey of Robotic Harvesting Systems and Enabling Technologies," *J. Intell. Robot. Syst.*, vol. 107, no. 2, p. 21, 2023.

[162] J. M. Brotons-Martínez, B. Martin-Gorriz, A. Torregrosa, and I. Porras, "Economic Evaluation of Mechanical Harvesting of Lemons," *Outlook Agric.*, vol. 47, no. 1, pp. 44–50, 2018.

[163] B. Mosadegh et al., "Pneumatic Networks for Soft Robotics That Actuate Rapidly," *Adv. Funct. Mater.*, vol. 24, no. 15, pp. 2163–2170, 2014.

[164] R. F. Shepherd et al., "Multigait Soft Robot," *Proc. Natl. Acad. Sci.*, vol. 108, no. 51, pp. 20400–20403, 2011.

[165] Y. Su et al., "A High-Payload Proprioceptive Hybrid Robotic Gripper with Soft Origamic Actuators," *IEEE Robot. Autom. Lett.*, vol. 5, no. 2, pp. 3003–3010, 2020.

[166] V. Marinoudi, C. G. Sørensen, S. Pearson, and D. Bochtis, "Robotics and Labour in Agriculture. A Context Consideration," *Biosyst. Eng.*, vol. 184, pp. 111–121, 2019.

[167] F. Vulpi, R. Marani, A. Petitti, G. Reina, and A. Milella, "An RGB-D Multi-View Perspective for Autonomous Agricultural Robots," *Comput. Electron. Agric.*, vol. 202, p. 107419, 2022.

[168] F. B. P. Malavazi, R. Guyonneau, J.-B. Fasquel, S. Lagrange, and F. Mercier, "LiDAR-Only Based Navigation Algorithm for an Autonomous Agricultural Robot," *Comput. Electron. Agric.*, vol. 154, pp. 71–79, 2018.

[169] T. Ji, A. N. Sivakumar, G. Chowdhary, and K. Driggs-Campbell, "Proactive Anomaly Detection for Robot Navigation with Multi-Sensor Fusion," *IEEE Robot. Autom. Lett.*, vol. 7, no. 2, pp. 4975–4982, 2022.

[170] Y. Yan, B. Zhang, J. Zhou, Y. Zhang, and X. Liu, "Real-Time Localization and Mapping Utilizing Multi-Sensor Fusion and Visual–IMU–Wheel Odometry for Agricultural Robots in Unstructured, Dynamic and GPS-Denied Greenhouse Environments," *Agronomy*, vol. 12, no. 8, 2022.

[171] Z. Zhang, P. Li, S. Zhao, Z. Lv, F. Du, and Y. An, "An Adaptive Vision Navigation Algorithm in Agricultural IoT System for Smart Agricultural Robots," *Comput. Mater. Contin.*, vol. 66, no. 1, 2021.

[172] R. R. Shamshiri and C. Weltzien, "Development and Field Evaluation of a Multichannel LoRa Sensor for IoT Monitoring in Berry Orchards," in *41. GIL-Jahrestagung, Informations-und Kommun. Krit. Zeiten*. Bonn: Gesellschaft für Informatik e.V. PISSN: 1617-5468. ISBN: 978-3-88579-703-6, GIL-Jahrestagung - Fokus: Informations- und Kommunikationstechnologien in kritischen Zeiten. Potsdam, Online. 8–9 März 2021, pp. 289–294. https://dl.gi.de/server/api/core/bitstreams/c082382d-29dd-4fcf-a558-c4a632fd5c68/content

[173] T. Li and D. Li, "Prospects for the Application of 5G Technology in Agriculture and Rural Areas," in *2020 5th International Conference on Mechanical, Control and Computer Engineering (ICMCCE)*, Harbin, China, 2020, pp. 2176–2179. https://doi.org/10.1109/ICMCCE51767.2020.00472

[174] G. Valecce, S. Strazzella, and L. A. Grieco, "On the Interplay Between 5G, Mobile Edge Computing and Robotics in Smart Agriculture Scenarios," in *Ad-Hoc, Mobile, and Wireless Networks. ADHOC-NOW 2019*. Lecture Notes in Computer Science, vol. 11803, M. Palattella, S. Scanzio, and S. Coleri Ergen, Eds. Cham: Springer, 2019, pp. 549–559. https://doi.org/10.1007/978-3-030-31831-4_38

[175] A. Khanna and S. Kaur, "Evolution of Internet of Things (IoT) and Its Significant Impact in the Field of Precision Agriculture," *Comput. Electron. Agric.*, vol. 157, pp. 218–231, 2019.

[176] Y. Tang, S. Dananjayan, C. Hou, Q. Guo, S. Luo, and Y. He, "A Survey on the 5G Network and Its Impact on Agriculture: Challenges and Opportunities," *Comput. Electron. Agric.*, vol. 180, p. 105895, 2021.

[177] R. Ma, K. H. Teo, S. Shinjo, K. Yamanaka, and P. M. Asbeck, "A GaN PA for 4G LTE-Advanced and 5G: Meeting the Telecommunication Needs of Various Vertical Sectors Including Automobiles, Robotics, Health Care, Factory Automation, Agriculture, Education, and More," *IEEE Microw. Mag.*, vol. 18, no. 7, pp. 77–85, 2017.

[178] M. van Hilten and S. Wolfert, "5G in Agri-Food—A Review on Current Status, Opportunities and Challenges," *Comput. Electron. Agric.*, vol. 201, p. 107291, 2022.

[179] H. Wang, Y. Yu, and Q. Yuan, "Application of Dijkstra Algorithm in Robot Path-Planning," in *2011 Second International Conference on Mechanic Automation and Control Engineering*, Hohhot, 2011, pp. 1067–1069. https://doi.org/10.1109/MACE.2011.5987118

[180] L.-B. Chen, X.-R. Huang, and W.-H. Chen, "Design and Implementation of an Artificial Intelligence of Things-Based Autonomous Mobile Robot System for Pitaya Harvesting," *IEEE Sens. J.*, p. 1, 2023.

5 Toward Optimizing Path Tracking of Agricultural Mobile Robots with Different Steering Mechanisms
A Simulation Framework

Redmond R. Shamshiri

5.1 INTRODUCTION

Waypoint-based path tracking [1] is essential for agricultural mobile robots to perform tasks such as field mapping [2], crop monitoring [3], autonomous spraying [4], or harvesting [5] operations. This involves guiding the robot along a predefined route by sequentially navigating through a series of specified locations called waypoints until all waypoints have been reached. Each waypoint represents a key position or landmark that the robot needs to reach during its mission. The robot's task is to autonomously navigate from one waypoint to the next while following a desired path. Since agricultural robots are typically battery-powered, energy efficiency is crucial for continued operation in the field, which requires implementation of optimized path-tracking algorithms that can minimize unnecessary movements and optimize routes to conserve energy while completing tasks effectively. In dynamic and unstructured agricultural environments, path tracking presents various challenges, such as finding the shortest path between several waypoints while driving on uneven terrain, slopes, bumps, and avoiding obstacles such as rocks or tree roots. At each waypoint, the robot determines the next waypoint based on a predefined route or mission plan. The path between waypoints can be predefined based on prior knowledge of the environment or generated dynamically in real time. This path may consist of straight lines, curves, or complex trajectories depending on the terrain and mission requirements. As the robot navigates, it continuously compares its current position with the coordinates of the next waypoint and adjusts its trajectory and control inputs to steer towards the target waypoint while avoiding obstacles and adhering to any specified constraints. To ensure accurate waypoint tracking, the robot typically employs feedback control mechanisms such as Proportional-Integral-Derivative (PID) control [6]

DOI: 10.1201/9781003306283-5

to continuously adjust the robot's steering and speed based on the difference between the desired path (defined by waypoints) and the actual path followed by the robot.

The objective of this study was to design a general simulation framework that incorporates MATLAB® (MathWorks Inc, MA, USA) programming and a CoppeliaSim environment (Coppelia Robotics AG, Zurich, Switzerland) in order to investigate and evaluate various path tracking control strategies for agricultural mobile robots. To this aim, an overview of the effect of different steering mechanisms on path tracking performance of agricultural mobile robots in different field conditions has been provided. Additionally, we aim to explore methods for finding the shortest path covering several random waypoints provided to the robot by the farmers. We present the design and implementation of a PID controller to regulate the steering angle and speed of a robot, enabling it to track a reasonable shortest path between these multiple randomly assigned waypoints. A sub-section has been provided to discuss the future work on the application and performance of a model predictive controller (MPC) in comparison to the PID controller for path tracking tasks. Through these investigations, our goal is to provide insights into the effectiveness of various path tracking control strategies and their applicability in enhancing the navigation capabilities of agricultural mobile robots.

5.2 EFFECT OF STEERING MECHANISM ON PATH TRACKING

The choice of steering mechanism significantly influences the robot's maneuverability, efficiency, and adaptability to different agricultural field terrain and tasks. Some of the most commonly employed steering systems in agricultural mobile robots include differential drive [7–9], Ackermann steering [10], articulated steering [11], and skid steering [12], each presenting different advantages and limitations that needs to be considered in selecting the most suitable option for a path tracking controller under specific field conditions.

Differential drive steering, characterized by independent control of the left and right wheels, offers a simple yet effective solution for agricultural robots, especially where precise positioning and maneuverability are required, such as orchards and vineyards. These robots typically consist of a chassis equipped with two independently driven wheels on either side that are often powered by DC motors or servomotors, enabling the robot to move and steer by varying the speeds of its left and right wheels independently. This steering mechanism facilitates precise control over the robot's movement, enabling it to navigate through rough terrain and narrow rows of crops and execute complex maneuvers with ease. Moreover, the simplicity of the differential drive system translates into lower manufacturing costs and easier maintenance, making it an attractive option for small to medium-sized farms seeking cost-effective automation solutions. They are perhaps the simplest and most common types of mobile robots used in agriculture for applications such as crop health assessment, field monitoring, and automated weeding.

Figure 5.1 shows several examples of agricultural wheeled robots with differential drive steering mechanisms, including two porotype models called Xf-Rovim [13] and RobHortic [14]. The Xf-Rovim, designed for early detection of *Xylella fastidiosa* (*X. fastidiosa*) in olive orchards, employs two DC motor control drivers,

FIGURE 5.1 Examples of differential steering robots showing (a) Xf-Rovim [13], (b) RobHortic [14], (c) Naïo Oz440 weeding robot (image source: futurefarming.com), (d) Fendt Xaver seeding robot (image source: profi.co.uk), (e) Fendt Xavier planting robot (image source: fendt.com), and (f) Mamut robot used for crop disease detection (image source: wevolver.com).

specifically H-Bridge modules. These modules enable the operator to have pre-cise control over the speed, torque, and direction of rotation for each wheel. The utilization of a dedicated driver for each wheel enhances maneuverability. This feature is particularly useful for reducing the rotation angle and achieving greater precision in the robot's movements. RobHortic is a field robot designed to detect the presence of pests and diseases in horticultural crops. The robot contained a frame with four fat-bike wheels designed to navigate through uneven terrains seamlessly. While the front wheels remain fixed, the two rear wheels, which are comparatively smaller, can rotate freely, enhancing the robot's adaptability to diverse field condi-tions. However, the limitations of differential drive steering become apparent when confronted with uneven terrain or obstacles. The inability to pivot around a central point limits its effectiveness in scenarios requiring tight turns or traversing rough terrain. Additionally, differential drive robots may struggle with stability on slopes or inclines, compromising their performance in hilly or uneven agricultural land-scapes. Table 5.1 lists a brief summary of some of the agricultural mobile robots with differential drive mechanisms.

In contrast to the simplicity of differential drive steering, Ackermann (and double-Ackerman) steering, as shown in Figure 5.2, offers a more sophisticated approach that prioritizes stability and smooth maneuverability. By coordinating the steering angles of all wheels to ensure they converge at a central point, Ackermann steering enables robots to execute precise turns while maintaining stability, even on uneven terrain. This makes Ackermann steering particularly well-suited for applications such as plowing, seeding, and towing in large-scale agricultural operations where stability

TABLE 5.1
Examples of Agricultural Mobile Robots with Differential Drive Mechanism

Robot	Description	Application/Use
TerraSentia [15]	Compact differential drive robot equipped with sensors for plant phenotyping and agricultural research	Crop monitoring, phenotyping, data collection
Ladybird [16]	Autonomous robotic platform designed for field scouting, weed detection, and precision spraying in agricultural fields	Weed detection, pest management, precision agriculture
Thorvald [17, 18]	Versatile robotic platform equipped with various modules for different agricultural tasks such as weeding and monitoring	Crop inspection, weeding, research
SwagBot [19]	Autonomous robot designed for monitoring and managing livestock in challenging terrain and environments	Livestock monitoring, herding, pasture management
FarmDroid FD20 [20]	Robotic seeding and planting machine capable of autonomously navigating fields and planting crops with precision.	Seeding, planting, precision agriculture

FIGURE 5.2 Examples of two-wheeled mobile robots used in agriculture with double-Ackerman steering, (a) Irus Quatrak (top image) [21] and (b) VIONA, built by Robot Makers GmbH (image source: robotmakers.de).

and efficiency are paramount. However, despite these advantages, Ackermann steering may pose challenges in navigating narrow passages or confined spaces due to its larger turning radius compared to differential drive systems. Moreover, the complexity of Ackermann steering mechanisms may result in higher manufacturing costs and maintenance requirements, limiting its accessibility to smaller-scale agricultural operations with more constrained budgets. Examples of Ackerman steering robots are the conventional autonomous tractors with two front passive and steerable wheels and two rear fixed and active wheels, or the four-wheel-drive double Ackerman steering robotic platform called VIONA (Vehicle for Intelligent Off-road NAvigation), which has been specifically designed and built for use in rough terrain by Robot Makers GmbH.

Articulated steering introduces a level of flexibility and maneuverability that is especially beneficial for large-scale agricultural operations encompassing vast fields and varying terrains. By articulating at a pivot point between two sections, articulated steering enables robots to navigate tight turns and negotiate uneven terrain with greater agility and precision. This makes articulated steering an ideal choice for tasks such as plowing, seeding, and harvesting in expansive agricultural landscapes where efficiency and adaptability are paramount. Despite its advantages, an articulated steering system may introduce complexities in design and operation, necessitating sophisticated control algorithms and robust mechanical components to ensure reliable performance. Moreover, the articulated joint presents a potential point of vulnerability to mechanical wear and tear, requiring regular maintenance and potentially increasing downtime. An example of an articulated steering heavy platform is the Weidemann Hoftrac 1190e electrical tractor designed primarily for agricultural applications. It has a length of approximately 3.7 m, width of approximately 1.8 m, height of 2.2 m, operating weight of around 3,500 kg, and maximum lifting loads of up to 1,500 kg. The tractor features an articulated steering system, allowing for greater maneuverability, especially in tight spaces common in agricultural environments, and benefits from all-wheel drive, which provides traction and stability on various terrains, including uneven or muddy surfaces typically encountered in agricultural fields. Figure 5.3 shows other examples of agricultural mobile robots with articulated steering mechanisms, including (a) an electric UGV called Tracdrone that benefits from a hydraulic drive, (b) a concept robot called the GEISI developed by Robot Makers GmbH that is equipped with a passive articulated steering system and single-wheel drive, and (c) the PhenoBot 3.0 wheeled agricultural robot with special machine-learning-based software that has been designed and developed by a team from North Carolina State University and Iowa State University, to navigate between corn rows spaced 0.76 m apart [22].

5.3 FINDING THE SHORTEST PATH

In mobile robotics, finding the shortest path is a fundamental problem that arises in various scenarios, ranging from navigation in structured environments to exploration

FIGURE 5.3 Examples of agricultural mobile robots with articulated steering mechanism, (a) *Tracdrone* (Image source: tracdrone.com), (b) GEISI (Image source: robotmakers.de), and (c) PhenoBot 3.0 [22], (image source: newatlas.com).

TABLE 5.2
List of Commonly Used Path Finding Methods in Mobile Robotics

Method	Description	Solution Guarantee	Main Limitations
Dijkstra's Algorithm [23]	Explores the graph from the starting node in a greedy manner, visiting the nearest unvisited node first and updating the shortest distances to reach each node.	Optimal solution for non-negative edge weights	Inefficient for graphs with negative edge weights or cycles, requires storing and updating distance values for all nodes
A* Search [24, 25]	Combines the advantages of Dijkstra's algorithm with heuristic information to guide the search towards the goal. It evaluates nodes based on the cost to reach them from the start node plus a heuristic estimate of the cost to reach the goal.	Optimal solution with admissible heuristics	Quality of solution depends on the heuristic used, not guaranteed to find optimal solution without an admissible heuristic
Bellman-Ford Algorithm [26]	Finds the shortest path by relaxing edges repeatedly and detects negative weight cycles. It iterates over all edges multiple times, updating the shortest distance to each node.	Optimal solution for graphs without negative weight cycles	Slower than Dijkstra's algorithm, especially on dense graphs, may not work correctly with negative weight cycles
Floyd-Warshall Algorithm [27]	Computes the shortest paths between all pairs of nodes in a weighted graph. It iteratively updates a matrix of shortest distances by considering all possible intermediate nodes.	Optimal solution for all pairs of nodes	Requires $O(V^3)$ time and space, not suitable for large graphs, may not work correctly with negative weight cycles
Bidirectional Search [25, 28]	Explores the graph simultaneously from both the start and goal nodes, meeting in the middle. It reduces the search space by focusing on paths from each direction towards the meeting point.	Optimal solution for unweighted or uniform cost graphs	Requires bidirectional edges and knowledge of the goal, may not always be applicable or efficient
Genetic Algorithm [29]	Mimics the process of natural selection and evolution to find the optimal solution by evolving a population of candidate solutions over multiple generations.	Heuristic-based approach, may find near-optimal solutions	Convergence to optimal solution not guaranteed, solution quality highly dependent on parameter tuning and representation encoding

(Continued)

TABLE 5.2 (Continued)
List of Commonly Used Path Finding Methods in Mobile Robotics

Method	Description	Solution Guarantee	Main Limitations
Simulated Annealing [30]	It is a probabilistic optimization algorithm inspired by the annealing process in metallurgy. It explores the solution space by accepting worse solutions with a decreasing probability, allowing it to escape local optima.	May find near-optimal solutions	Convergence to optimal solution not guaranteed, performance sensitive to parameter tuning and cooling schedule
Ant Colony Optimization [31]	It is inspired by the foraging behavior of ants. It involves simulating a colony of ants laying pheromone trails on paths, with pheromone levels influencing the probability of path selection by subsequent ants.	Heuristic-based approach, may find near-optimal solutions	Convergence to optimal solution not guaranteed, highly sensitive to parameter settings and pheromone update rules
Nearest Neighbor [32]	It starts from an arbitrary point and repeatedly selects the closest unvisited point until all points are visited, forming a tour.	Greedy algorithm, suboptimal solution	Does not guarantee an optimal solution, highly dependent on the starting point
Christofides Algorithm [REF]	It combines a minimum spanning tree with a matching algorithm to find a tour that is guaranteed to be within a factor of 3/2 of the optimal solution for the traveling salesman problem.	Near-optimal solution	More complex implementation compared to other methods, not always practical for small instances due to overhead
Dynamic Programming [33, 34]	Solves TSP by breaking down the problem into smaller subproblems and solving them recursively. It finds the optimal solution but is computationally expensive.	Optimal solution for small instances	Exponential time complexity, impractical for large instances
Branch and Bound [35–37]	Enumerates all possible permutations of cities and prunes branches that cannot lead to an optimal solution. Guarantees an optimal solution for small instances but is computationally expensive.	Optimal solution for small instances	Exponential time complexity, memory-intensive for large instances

in unknown terrains. Agricultural mobile robots are sometimes required to cover the entire field while minimizing the distance traveled to complete a specific task efficiently. Optimizing path planning by finding the shortest path can contribute to minimizing the energy consumption of the robot, saving operational time and reducing crop damage and potential environmental impacts (i.e., reduce soil compaction and minimize disruption to crops, preserving soil health and biodiversity in the long term). Additionally, finding the shortest path is sometimes essential for autonomous navigation in challenging terrain or complex environments that may contain obstacles such as trees, irrigation systems, or uneven terrain, which can pose risks to the robot. One commonly employed approach in mobile robotics is the use of graph-based algorithms that represent the environment as a graph, with nodes representing locations and edges representing possible transitions between these locations. By modeling the environment in this way, robots can apply graph search algorithms to find the shortest path from their current position to a target location. Some of the most widely used methods and algorithms for finding the shortest path used in mobile robots are listed in Table 5.2. Two of the most popular graph search algorithms are Dijkstra's algorithm, which explores the graph from the robot's current position outward, considering the cost of reaching neighboring nodes and updating the shortest path accordingly, and the A* search, which incorporates heuristic information to guide the search towards the goal more efficiently, often resulting in faster convergence to the optimal path. In addition to these classic algorithms, mobile robots often leverage specialized techniques custom-designed to their specific requirements and constraints. For example, in dynamic environments where obstacles can move or appear unpredictably, algorithms such as Rapidly-exploring Random Trees (RRT) or its variants such as RRT* are preferred for their ability to quickly adapt to changing conditions and find feasible paths in high-dimensional spaces.

In this study, a heuristic approach [38] was used to find the shortest path (or a reasonably short path) by using the Nearest Neighbor (NN) algorithm [32, 39] which can find the optimal or near-optimal solution. This path will be then sent to the robot navigation system manually or by means of a wireless teleoperation system. Given the set of WP = $\{wp_1, wp_2,. . ., wp_n\}$ waypoints with coordinates (x_i, y_i) for $i = 1,2, . . .,$ $n \in \mathbb{N}$, where n is the total number of waypoints in the path, the shortest possible path for the robot to visit each waypoint exactly once and returns to the starting waypoint, was calculated by letting the path [wp_0] be the tour path, where wp_0 is an arbitrary chosen starting waypoint from WP. The distance between two consecutive waypoints wp_i and wp_j is calculated using the Euclidean distance given by Equation 5.1. At each step, the algorithm selects the nearest unvisited waypoint wp_i to the last waypoint wp_{i-1} in the path by finding the minimum d_i, and adds that selected nearest waypoint wp_i to the path until all waypoints are visited. This method provided a quick solution that can yield the shortest path (or a reasonably short path) with a total path length of $\sum^{n-1} d_i$. However, it should be noted that the NN algorithm may produce a path that is significantly longer than the optimal shortest path, particularly when the robot is dealing with a large number of WPs with complex arrangements. To find the absolute shortest path, other algorithms (i.e., dynamic programming, branch and bound, or genetic algorithms) are required to find the optimal solution, which is resource-intensive. Figure 5.4 demonstrates the results of employing NN algorithm on a randomly selected set of waypoints with size of 5, 15, 50, and 100. The MATLAB code used to implement the NN algorithm and generate the result is provided in Appendix 1.

$$d_i(wp_{i-1}, wp_i) = \sqrt{(x_{i-1} - x_i)^2 + (y_{i-1} - y_i)^2} \qquad (5.1)$$

It should be noted that with the advancement of artificial intelligence and machine learning, mobile robots can now employ learning-based approaches for path planning, such as reinforcement learning algorithms that enable robots to learn optimal navigation policies through trial and error, taking into account environmental dynamics and task-specific objectives. However, despite the diversity of approaches, challenges persist in real-world robotics applications, including the need to handle complex environments with dynamic obstacles, ensuring real-time performance on resource-constrained platforms, and addressing uncertainties inherent in sensor measurements and actuator responses.

5.4 PATH TRACKING CONTROLLERS

During actual field operation with uncertain environments and variations in terrain, mobile robots often exhibit nonlinear dynamics, making the design of effective navigation controllers challenging. Sensors such as odometry, GPS, LiDAR, or cameras may suffer from noise, biases, or measurement inaccuracies, affecting the accuracy of path tracking. Therefore, path tracking controllers must operate within real-time constraints to ensure timely responses to these changing conditions. The primary objective of a path tracking controller is to minimize the deviation between the robot's actual trajectory and the desired path. The choice

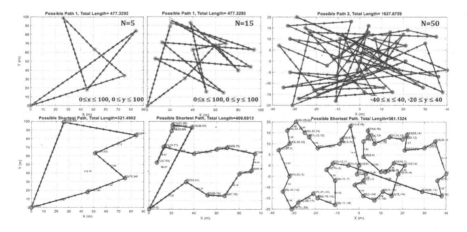

FIGURE 5.4 Plots of random path and possible shortest path connecting N points calculated using the nearest neighbor algorithm for providing agricultural mobile robots with a reasonably short path.

of controller depends on factors such as the robot's dynamics, sensor capabilities, environmental conditions, and performance requirements. Controllers such as PID, Stanley, Model Predictive Control (MPC) [11], Sliding Mode Control (SMC) [40], Pure Pursuit [41, 42], Linear Quadratic Regulator (LQR) [43], Artificial Neural Networks (ANNs) [44], Fuzzy Logic Control [45], and Reinforcement Learning (RL) [46] have been tested in the navigation systems of agricultural mobile robots for following predefined waypoints and trajectories in various environments, ranging from structured indoor fields to challenging outdoor terrains. To enhance path tracking performance, state estimation techniques such as Kalman filtering [47] or observer design are also integrated to enable the estimation of unmeasured states, providing valuable insights into the robot's position, orientation, and velocity. Accurate state estimation leads to precise control actions based on comprehensive system information, resulting in improved overall path tracking performance. Some studies have benefited from a robust control design such as H-infinity control that offer robust stability and performance across a spectrum of operating conditions for ensuring reliable performance in the face of uncertainties or disturbances encountered in agricultural environments. Nevertheless, experimentation and validation in simulation environments and real-world agricultural settings are essential for assessing the effectiveness and applicability of these control methods in practical agricultural robotics applications.

5.4.1 PID CONTROLLER

The differential drive mobile robot with four wheels shown in Figure 5.5 was used for path tracking simulation purposes and implementing PID controllers. The control strategy employed in the simulation is aimed at guiding a robot along a predefined path while regulating its speed. The steering angle controller provides the robot with the autonomy to align itself with the next waypoint, incorporating P, I, and D terms to minimize orientation errors. Similarly, the speed controller adjust the robot's linear velocity to maintain a desired speed profile as it approaches waypoints. Both controllers employ feedback mechanisms, continually recalculating control inputs based on the current state of the robot and the proximity to the target waypoint. The simulation loop iterates over a predefined number of time steps, during which the robot continuously updates its position, orientation, and control inputs based on the feedback received from the environment. At each time step, the controller computes the new steering angle and linear velocity using the PID controllers, and enforces constrains, ensuring the robot's motion remains within safe operational bounds and performs smooth and accurate navigation towards the waypoints.

The kinematic equations given in Equations 5.2–5.4 alongside two PID controllers given in Equations 5.5–5.6 were implemented in MATLAB and CoppeliaSim in order to propose a general simulation framework that can be used for experimenting with path tracking and speed control of agricultural mobile robots and simulate their behavior across a spectrum of real-world scenarios with any sequence of waypoints. Here we assume the mobile robot shown in Figure 5.5, which has a maximum and minimum linear velocity of v_{max} and v_{min}, angular velocity of ω, wheelbase of L, wheel radius of R, maximum allowable steering angle of δ_{max}, and that the robot is currently located at point (x_t, y_t) with current orientation θ_t. The controller must then guide the robot to the next desired waypoint at (x_d, y_d) by providing it with two control signals, steering $\delta(t)$, and speed $\vartheta(t)$. The controller first calculates the desired orientation θ_d that the robot needs to have in order to align it with the line connecting its current position (x_t, y_t) to the next desired waypoint (x_d, y_d) and generates an error signal $\acute{e}\theta(t)$ (calculated using Equation 5.7) by taking into account its current orientation θ_t. This error is adjusted to ensure it falls within the range $-\pi$ to π, resulting in the final error signal $e(t)$, as described in Equation 5.8. The controller also calculates the integral of this error as $\int_0^t e_\theta(\tau)d\tau$ to θ 0 θ track the accumulated error over time, and the derivative of this error as $\frac{de(t)}{dt}$ to determine the rate of d_t change of the error signal and generate steering control $\delta(t)$ that influences the robot's trajectory towards the target point. A similar approach is used for the speed control by comparing the robot distant to the next waypoint with a reference distance $dist_{ref}$, resulting in a speed error signal of $e_v(t)$ (given by Equation 5.8). Finally, in each time step of the simulation, the change in the x-y coordinate of the robot's position over a small time interval Δt is calculated using Equation 5.2.

$$[x_{t+1}, y_{t+1}] = [x_t + v_t \cos(\theta_t) \cdot \Delta t, y_t + v_t \sin(\theta_t) \cdot \Delta t] \qquad (5.2)$$

$$\omega_t = \frac{v}{L} \tan(\delta(t)) = \frac{\theta_{t+1} - \theta_t}{\Delta t} \qquad (5.3)$$

$$\theta_{t+1} = \theta_t + (\frac{v}{L}) \cdot \tan(\omega_t) \cdot \Delta t \qquad (5.4)$$

$$\delta(t) = K_p \cdot e_\theta(t) + K_i \cdot \int_0^t e_\theta(\tau)d\tau + K_d \frac{de_\theta(t)}{dt} \qquad (5.5)$$

$$u_v(t) = Kp_v \cdot e_v(t) + Ki_v \cdot \int_0^t e_v(\tau)d\tau + Kd_v \frac{de(t)}{dt} \qquad (5.6)$$

$$\acute{e}_\theta(t) = \theta_d - \theta_t \qquad (5.7)$$

$$e_\theta(t) = atan2(\sin(\acute{e}_\theta(t)), \cos(\acute{e}_\theta(t))) \qquad (5.8)$$

$$e_v(t) = \sqrt{(x_d - x)^2 + (y_d - y)^2 - dist_{ref}} \qquad (5.9)$$

The gains of the PID controller for path tracking was tuned using a trial and error approach, as shown in Figure 5.6. The simulation framework was then integrated to provide visual feedback in the form of plots, illustrating the robot's trajectory, steering angle, angular velocity, linear velocity, orientation, and error metrics over time, as shown in Figure 5.7. These plots are used for better understanding of the path tracking controller, allowing real-time monitoring of

FIGURE 5.5 The differential drive agricultural mobile robot used in the CoppeliaSim and Sketchup environment for path tracking simulation.

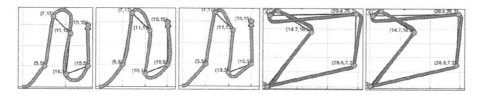

FIGURE 5.6 Experiments with tuning PID gains of the designed path tracking controller.

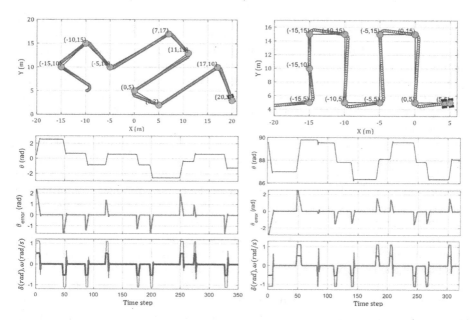

FIGURE 5.7 Results of PID controller for tracking two different paths, showing plots of the robot orientation (θ), orientation error (θ_{error}), steering command (δ), and angular velocity (ω).

the robot's behavior. In addition, various data points, such as the robot's path, angular velocity, steering angle, linear velocity, orientation, and error metrics, are also recorded and saved for post-simulation evaluation, analysis, and refinement of the control algorithms. The MATLAB code used to implement the PID path tracking controller and generate the result is provided in Appendix 2. To enhance controller performance, further experiments and fine-tuning are required for adjusting the PID gains to ensure a more optimal response, stability, and robustness in handling errors. In situations when the integral term accumulates large errors, the robot-driven path leads to overshoot or instability. To prevent this, implementing an anti-windup mechanism such as clamping or resetting the integral term can contribute to smoother control action. To anticipate and compensate for known disturbances or dynamics in the system, feedforward control can be implemented to improve tracking performance by proactively adjusting the control inputs based on the expected behavior of the system. In scenarios with uncertain or time-varying parameters, adaptive control techniques offer further improvement by adjusting controller parameters based on changing system dynamics or operating conditions.

5.4.2 MODEL PREDICTIVE CONTROLLER

Nonlinear control methods, such as model predictive control (MPC) or sliding mode control (SMC), provide sophisticated strategies for systems with nonlinear dynamics or constraints. These methods optimize control actions based on system models, offering superior performance in complex scenarios. To design an MPC controller for steering angle of the robot, a lateral dynamics of the robot given by the state space equation in (10) is used to predict the future behavior of the robot variables and optimize the control commands. Here Vx is the longitudinal velocity at the center of gravity of the robot, m is the total mass, Iz is the yaw moment of inertia of the robot, lf and lr are respectively the longitudinal distance from center of gravity to front and rear tires, C_{af} and C_{ar} are respectively the cornering stiffness of front and rear tires, δ is the steering angle, y is the lateral position, and θ is the robot orientation (yaw angle). It is expected that the predictive nature of MPC contributes to the robot path tracking in the presence of a large number of waypoints.

The state of the robot is then defined as $X = [x, y, \theta]$, with the control input as the steering angle δ, and $\dot{Y} = V_x\,\theta + V_y$. The objective of the controller is to minimize the error between the current position of the robot (x_t, y_t) and the reference trajectory that is generated from the set of desired waypoints (x_d, y_d). We can define a cost function given in Equation 5.11 to minimize this error, as well as penalties on control inputs and state deviations. Here N is the prediction horizon, (x_i, y_i) are the predicted positions at each time step, and R and Q are turning parameters to adjust the penalties on control inputs and state deviations. The controller then uses the kinematic equations in (5.12) and (5.13) to predict the future states of the robot over the prediction horizon, where v_i, θ_i, and δ_i are the linear velocity, orientation, and steering angle (control input) at time step i. The optimization problem can be solved at each time step to find the optimal control inputs that minimize the cost function subject to constraints. The predicted path with the smallest cost function is considered the optimal steering angle solution. This can be achieved using numerical optimization techniques such as gradient-based or gradient-free optimization methods. The simulation applies the first control input from the optimal solution to the system and repeats the process at the next time step. Adjusting the parameters R, Q, and prediction horizon N, and constraints can help fine-tune the controller's performance based on specific requirements and system dynamics. The preliminary results of the robot path tracking using MPC controller that aims at visiting 25 and 100 waypoints are shown in Figures 5.8 and 5.9, demonstrating a more precise performance specially in complex situations.

$$\frac{d}{dt}\begin{bmatrix} \dot{y} \\ \theta \\ \dot{\theta} \end{bmatrix} = \begin{bmatrix} -\dfrac{2C_{\alpha f}+2C_{\alpha r}}{mV_x} & 0 & -V_x-\dfrac{2C_{\alpha f}+2C_{\alpha r}l_r}{mV_x} \\ 0 & 0 & 1 \\ -\dfrac{2l_fC_{\alpha f}+2C_{\alpha r}l_r}{I_zV_x} & 0 & -\dfrac{2L_f^2C_{\alpha f}+2C_{\alpha r}l_r^2}{I_zV_x} \end{bmatrix}\begin{bmatrix} \dot{y} \\ \theta \\ \dot{\theta} \end{bmatrix} + \begin{bmatrix} \dfrac{2C_{\alpha f}}{m} \\ 0 \\ \dfrac{2C_{\alpha f}l_f}{I_z} \end{bmatrix}\delta \quad (5.10)$$

$$J = \sum_{i=1}^{N}\left((x_i-x_d)^2+(y_i-y_d)^2+R.\delta_i^2+Q.(x_i-x_{i-1})^2+Q.(y_i-y_{i-1})^2\right) \quad (5.11)$$

$$\begin{bmatrix} x_{i+1}, y_{i+1} \end{bmatrix} = \begin{bmatrix} x_i+v_i.\cos(\theta_i).\Delta t, \ y_i+v_i.\sin(\theta_i).\Delta t \end{bmatrix} \quad (5.12)$$

$$\theta_{i+1} = \theta_i + \left(\frac{v_i}{L}\right).\tan(\delta_i).\Delta t \quad (5.13)$$

5.5 CONCLUSION

This chapter presented an overview of path tracking control strategies for agricultural mobile robots with a focus on calculations of the shortest path between several random waypoints in the field. Through a review of literature and simulations results that were generated in MATLAB and CoppeliaSim, the effectiveness of differential drive, Ackermann steering, and articulated steering mechanisms were discussed, considering their advantages and limitations under different agricultural field conditions. The design and implementation of a PID controller for path tracking was also presented, demonstrating its capability to regulate steering angle and speed to achieve efficient navigation between waypoints. Future research directions, such as the application of MPC control for path tracking tasks for complex scenarios involving a large number of waypoints, were then outlined to further enhance the navigation capabilities of agricultral mobile robots. This exploration of advanced control methods promises to address challenges associated with nonlinear dynamics and uncertainties in real-world agricultural

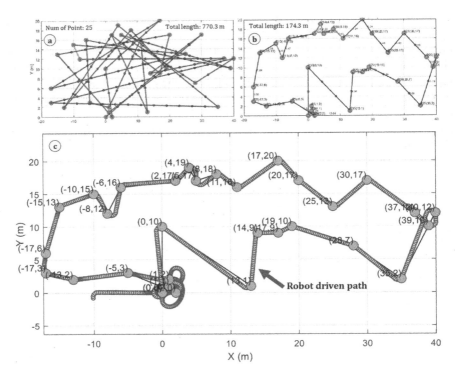

FIGURE 5.8 Demonstration of the (a) possible path connecting 25 waypoints with a total length of 770.3 m, (b) a possibly shortest path calculated by the robot controller with a total length of 174.3 m, and (c) the robot-driven trajectory using an MPC path tracking controller.

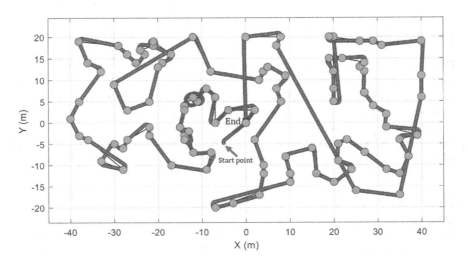

FIGURE 5.9 Trajectory of the robot performing path tracking using MPC controller with 100 waypoints.

environments, offering potential improvements in path tracking performance and robustness. It is expected that, with the aid of the proposed simulation framework, an optimized path tracking control strategy for agricultural mobile robots with different dimensions and steering mechanisms can be suggested in order to navigate within dynamically and unstructured agricultural environments with greater precision and efficiency.

REFERENCES

[1]   H. Kivrak, F. Cakmak, H. Kose, and S. Yavuz, "Waypoint Based Path Planner for Socially Aware Robot Navigation," *Cluster Computing*, vol. 25, no. 3, pp. 1665–1675, 2022.

[2]   D. Tiozzo Fasiolo, L. Scalera, E. Maset, and A. Gasparetto, "Towards Autonomous Mapping in Agriculture: A Review of Supportive Technologies for Ground Robotics," *Robotics and Autonomous Systems*, vol. 169, p. 104514, 2023.

[3]   N. Virlet, K. Sabermanesh, P. Sadeghi-Tehran, and M. J. Hawkesford, "Field Scanalyzer: An Automated Robotic Field Phenotyping Platform for Detailed Crop Monitoring," *Functional Plant Biology*, vol. 44, no. 1, pp. 143–153, 2017.

[4]   A. T. Meshram, A. V. Vanalkar, K. B. Kalambe, and A. M. Badar, "Pesticide Spraying Robot for Precision Agriculture: A Categorical Literature Review and Future Trends," *Journal of Field Robotics*, vol. 39, no. 2, pp. 153–171, 2022.

[5]   C. J. Hohimer, H. Wang, S. Bhusal, J. Miller, C. Mo, and M. Karkee, "Design and Field Evaluation of a Robotic Apple Harvesting System with a 3D-Printed Soft-Robotic End-Effector," *Transactions of the ASABE*, vol. 62, no. 2, pp. 405–414, 2019.

[6]   J. He et al., "Path Tracking Control Method and Performance Test Based on Agricultural Machinery Pose Correction," *Computers and Electronics in Agriculture*, vol. 200, p. 107185, 2022.

[7]   J. Iqbal, R. Xu, H. Halloran, and C. Li, "Development of a Multi-Purpose Autonomous Differential Drive Mobile Robot for Plant Phenotyping and Soil Sensing," *Electronics*, vol. 9, no. 9, 2020.

[8]   P. Huang, Z. Zhang, X. Luo, J. Zhang, and P. Huang, "Path Tracking Control of a Differential-Drive Tracked Robot Based on Look-ahead Distance," *IFAC-PapersOnLine*, vol. 51, no. 17, pp. 112–117, 2018.

[9]   G. Monsalve, N. B. Ltaief, V. Amoriya, and A. Cardenas, "Kinematic Navigation Control of Differential Drive Agricultural Robot," *2022 IEEE International Conference on Electrical Sciences and Technologies in Maghreb (CISTEM)*, vol. 4, pp. 1–6, 2022.

[10] R. F. Carpio et al., "A Navigation Architecture for Ackermann Vehicles in Precision Farming," *IEEE Robotics and Automation Letters*, vol. 5, no. 2, pp. 1103–1110, 2020.

[11] B. Zhou, X. Su, H. Yu, W. Guo, and Q. Zhang, "Research on Path Tracking of Articulated Steering Tractor Based on Modified Model Predictive Control," *Agriculture*, vol. 13, no. 4, 2023.

[12] C. Wen Zhu, E. Hill, M. Biglarbegian, S. A. Gadsden, and J. A. Cline, "Smart Agriculture: Development of a Skid-Steer Autonomous Robot with Advanced Model Predictive Controllers," *Robotics and Autonomous Systems*, vol. 162, p. 104364, 2023.

[13] B. Rey, N. Aleixos, S. Cubero, and J. Blasco, "Xf-Rovim. A Field Robot to Detect Olive Trees Infected by Xylella Fastidiosa Using Proximal Sensing," *Remote Sensing*, vol. 11, no. 3, 2019.

[14] S. Cubero, E. Marco-Noales, N. Aleixos, S. Barbé, and J. Blasco, "RobHortic: A Field Robot to Detect Pests and Diseases in Horticultural Crops by Proximal Sensing," *Agriculture*, vol. 10, no. 7, 2020.

[15] E. Kayacan, Z.-Z. Zhang, and G. Chowdhary, "Embedded High Precision Control and Corn Stand Counting Algorithms for an Ultra-Compact 3D Printed Field Robot," *Robotics: Science and Systems*, vol. 14, p. 9, 2018.

[16] A. Bender, B. Whelan, and S. Sukkarieh, "A High-Resolution, Multimodal Data Set for Agricultural Robotics: A Ladybird's-eye View of Brassica," *Journal of Field Robotics*, vol. 37, no. 1, pp. 73–96, 2020.

[17] L. Grimstad and P. J. From, "The Thorvald II Agricultural Robotic System," *Robotics*, vol. 6, no. 4, 2017.

[18] L. Grimstad and P. J. From, "Software Components of the Thorvald II Modular Robot," *Modeling, Identification and Control*, vol. 39, no. 3, pp. 157–165, 2018.

[19] S. Eiffert, N. D. Wallace, H. Kong, N. Pirmarzdashti, and S. Sukkarieh, "Experimental Evaluation of a Hierarchical Operating Framework for Ground Robots in Agriculture," in *Experimental Robotics. ISER 2020*. Springer Proceedings in Advanced Robotics, vol. 19, B. Siciliano, C. Laschi, and O. Khatib, Eds. Cham: Springer, 2021, pp. 151–160. https://doi.org/10.1007/978-3-030-71151-1_14

[20] R. Gerhards, P. Risser, M. Spaeth, M. Saile, and G. Peteinatos, "A Comparison of Seven Innovative Robotic Weeding Systems and Reference Herbicide Strategies in Sugar Beet (Beta vulgaris subsp. vulgaris L.) and Rapeseed (Brassica napus L.)," *Weed Research*, vol. 64, no. 1, pp. 42–53, 2024.

[21] R. Shamshiri, C. Weltzien, I. Zytoon, and B. Sakal, "Evaluation of Laser and Infrared Sensors with CANBUS Communication for Collision Avoidance of a Mobile Robot," in *Proceedings International Conference on Agricultural Engineering. AgEng-LAND. TECHNIK 2022*, V. D. I. Wissensforum, Ed. Düsseldorf: VDI Verlag GmbH (0083-5560/978-3-18092406-9), 2022, pp. 121–130. https://www.vdi-nachrichten.com/shop/ageng-land-technik-2022/

[22] L. Xiang, J. Gai, Y. Bao, J. Yu, P. S. Schnable, and L. Tang, "Field-Based Robotic Leaf Angle Detection and Characterization of Maize Plants Using Stereo Vision and Deep Convolutional Neural Networks," *Journal of Field Robotics*, vol. 40, no. 5, pp. 1034–1053, 2023.

[23] M. R. Wayahdi, S. H. N. Ginting, and D. Syahputra, "Greedy, A-Star, and Dijkstra's Algorithms in Finding Shortest Path," *International Journal of Advances in Data and Information Systems*, vol. 2, no. 1 SE, pp. 45–52, 2021.

[24] Y. Yang et al., "An Optimal Goal Point Determination Algorithm for Automatic Navigation of Agricultural Machinery: Improving the Tracking Accuracy of the Pure Pursuit Algorithm," *Computers and Electronics in Agriculture*, vol. 194, p. 106760, 2022.

[25] C. Li, X. Huang, J. Ding, K. Song, and S. Lu, "Global Path Planning Based on a Bidirectional Alternating Search A* Algorithm for Mobile Robots," *Computers & Industrial Engineering*, vol. 168, p. 108123, 2022.

[26] M. Parimala, S. Broumi, K. Prakash, and S. Topal, "Bellman–Ford Algorithm for Solving Shortest Path Problem of a Network Under Picture Fuzzy Environment," *Complex & Intelligent Systems*, vol. 7, no. 5, pp. 2373–2381, 2021.

[27] D. Lyu, Z. Chen, Z. Cai, and S. Piao, "Robot Path Planning by Leveraging the Graph-Encoded Floyd Algorithm," *Future Generation Computer Systems*, vol. 122, pp. 204–208, 2021.

[28] J. Wang, B. Li, and M. Q.-H. Meng, "Kinematic Constrained Bi-directional RRT with Efficient Branch Pruning for Robot Path Planning," *Expert Systems with Applications*, vol. 170, p. 114541, 2021.

[29] M. A. Yakoubi and M. T. Laskri, "The Path Planning of Cleaner Robot for Coverage Region using Genetic Algorithms," *Journal of Innovation in Digital Ecosystems*, vol. 3, no. 1, pp. 37–43, 2016.

[30] W. Shi, Z. He, W. Tang, W. Liu, and Z. Ma, "Path Planning of Multi-Robot Systems with Boolean Specifications Based on Simulated Annealing," *IEEE Robotics and Automation Letters*, vol. 7, no. 3, pp. 6091–6098, 2022.

[31] L. Wang, J. Kan, J. Guo, and C. Wang, "3D Path Planning for the Ground Robot with Improved Ant Colony Optimization," *Sensors*, vol. 19, no. 4, 2019.

[32] B. Ning, Q.-L. Han, Z. Zuo, J. Jin, and J. Zheng, "Collective Behaviors of Mobile Robots Beyond the Nearest Neighbor Rules with Switching Topology," *IEEE Transactions on Cybernetics*, vol. 48, no. 5, pp. 1577–1590, 2018.

[33] M. Ono, M. Pavone, Y. Kuwata, and J. Balaram, "Chance-Constrained Dynamic Programming with Application to Risk-Aware Robotic Space Exploration," *Autonomous Robots*, vol. 39, no. 4, pp. 555–571, 2015.

[34] A. Ayari and S. Bouamama, "A New Multiple Robot Path Planning Algorithm: Dynamic Distributed Particle Swarm Optimization," *Robotics and Biomimetics*, vol. 4, no. 1, p. 8, 2017.

[35] L. Bukata, P. Šůcha, and Z. Hanzálek, "Optimizing Energy Consumption of Robotic Cells by a Branch & Bound Algorithm," *Computers & Operations Research*, vol. 102, pp. 52–66, 2019.

[36] J. M. Palacios-Gasós, E. Montijano, C. Sagues, and S. Llorente, "Multi-robot Persistent Coverage Using Branch and Bound," in *2016 American Control Conference (ACC)*, Boston, MA, 2016, pp. 5697–5702. https://doi.org/10.1109/ACC.2016.7526562

[37] J. G. Martin, J. R. D. Frejo, R. A. García, and E. F. Camacho, "Multi-Robot Task Allocation Problem with Multiple Nonlinear Criteria Using Branch and Bound and Genetic Algorithms," *Intelligent Service Robotics*, vol. 14, no. 5, pp. 707–727, 2021.

[38] C. S. Tan, R. Mohd-Mokhtar, and M. R. Arshad, "A Comprehensive Review of Coverage Path Planning in Robotics Using Classical and Heuristic Algorithms," *IEEE Access*, vol. 9, pp. 119310–119342, 2021.

[39] M. Liu, "Robotic Online Path Planning on Point Cloud," *IEEE Transactions on Cybernetics*, vol. 46, no. 5, pp. 1217–1228, 2016.

[40] Y. An, L. Wang, X. Deng, H. Chen, Z. Lu, and T. Wang, "Research on Differential Steering Dynamics Control of Four-Wheel Independent Drive Electric Tractor," *Agriculture*, vol. 13, no. 9, 2023.

[41] L. Xu, Y. Yang, Q. Chen, F. Fu, B. Yang, and L. Yao, "Path Tracking of a 4WIS–4WID Agricultural Machinery Based on Variable Look-Ahead Distance," *Applied Sciences*, vol. 12, no. 17, 2022.

[42] M. Samuel, M. Hussein, and M. B. Mohamad, "A Review of Some Pure-Pursuit Based Path Tracking Techniques for Control of Autonomous Vehicle," *International Journal of Computer Applications*, vol. 135, no. 1, pp. 35–38, 2016.

[43] H. Chen, X. Wang, L. Zhao, R. Jiang, and B. Zhumadil, "Research on Path Tracking Control of Mobile Storage Robot Based on Model Predictive Control and Linear Quadratic Regulator," *Proc. SPIE*, vol. 12748, p. 1727482O, 2023.

[44] F. Vulpi, A. Milella, R. Marani, and G. Reina, "Recurrent and Convolutional Neural Networks for Deep Terrain Classification by Autonomous Robots," *Journal of Terramechanics*, vol. 96, pp. 119–131, 2021.

[45] Y. Wang, X. Li, J. Zhang, S. Li, Z. Xu, and X. Zhou, "Review of Wheeled Mobile Robot Collision Avoidance Under Unknown Environment," *Science Progress*, vol. 104, no. 3, p. 00368504211037771, 2021.

[46] L. Chang, L. Shan, C. Jiang, and Y. Dai, "Reinforcement Based Mobile Robot Path Planning with Improved Dynamic Window Approach in Unknown Environment," *Autonomous Robots*, vol. 45, no. 1, pp. 51–76, 2021.

[47] J. N. Greenberg and X. Tan, "Dynamic Optical Localization of a Mobile Robot Using Kalman Filtering-Based Position Prediction," *IEEE/ASME Transactions on Mechatronics*, vol. 25, no. 5, pp. 2483–2492, 2020.

Appendix 1: MATLAB® Code for Finding the Possible Shortest Path Connecting N Random Points Using the Nearest Neighbor Algorithm (A Solution to Solve the Travelling Salesman Problem Approximately)

```
% -------------------------------------------------------------------
% Code title:         Finding the possible shortest path
%                     connecting N points
% Author:             Redmond R. Shamshiri©, redmond@
%                     AdaptiveAgroTech.com
% Last update date:   March.01.2024
% -------------------------------------------------------------------

num_points = 5;NumSteps = 5;
x0 = 0; y0 = 0;
x = [x0 randi([0, 100], 1, num_points)]; y = [y0 randi
   ([0, 100], 1, num_points)];
original_x = x; original_y = y;
distances = pdist2([x', y'], [x', y']);
tsp_path = nearestNeighbor(distances);
figure('Color', 'white');
WP_Path = zeros(num_points + 2, 2);
for k = 1:5
  subplot(2, 3, k);
  idx = randperm(num_points) + 1;
  idx = [1, idx, 1];

  plot(original_x(idx), original_y(idx), 'Color', [0.6 0.7 0.8],
'LineWidth', 4, 'MarkerSize', 5, 'MarkerEdgeColor', 'black');
  hold on;
```

```
scatter(original_x(idx), original_y(idx), 100, 'w', 'filled',
  'MarkerEdgeColor', 'k'); title(['Possible Path ',
  num2str(k)]); xlabel('X (m)'); ylabel('Y (m)'); grid on;

  eval(['WP_Path', num2str(k), ' = zeros(num_points + 2, 2);']);
  eval(['WP_Path', num2str(k), '(:, 1) = original_x(idx);']);
  eval(['WP_Path', num2str(k), '(:, 2) = original_y(idx);']);

  for i = 1:length(idx)-1
    len = norm([original_x(idx(i+1))-original_x(idx(i)),
original_y(idx(i+1))-original_y(idx(i))]);
    steps = ceil(len / NumSteps);
    x_segment = linspace(original_x(idx(i)), original_x(idx(i+1)),
      steps);
    y_segment = linspace(original_y(idx(i)), original_y(idx
      (i+1)), steps);

    for j = 1:steps
      plot(x_segment(1:j), y_segment(1:j), 'Color', [0.1 0.1
        0.1], 'LineWidth', 1); hold on;
      plot(x_segment(j), y_segment(j), 'ro-', 'MarkerFaceColor',
'r', 'MarkerSize', 5, 'MarkerEdgeColor', 'k');
      pause(0.05);
    end

  end

  WP_Path(:, 1) = original_x(idx);
  WP_Path(:, 2) = original_y(idx);
end
subplot(2, 3, 6);
plot(original_x(tsp_path), original_y(tsp_path), 'r-o',
  'MarkerFaceColor', 'g', 'LineWidth', 2, 'MarkerSize', 10);
hold on;
for i = 1:num_points + 1
  text(original_x(tsp_path(i)), original_y(tsp_path(i)),
['P', num2str(i), '(', num2str(original_x(tsp_path(i))), ',',
num2str(original_y(tsp_path(i))), ')'], 'HorizontalAlignment',
'left', 'VerticalAlignment', 'bottom', 'FontSize', 8);
end
title('Possible Shortest Path'); xlabel('X (m)'); ylabel
  ('Y (m)'); grid on;
for i = 1:num_points
  j = mod(i, num_points) + 1;
  len = norm([original_x(tsp_path(j))-original_x(tsp_path(i)),
original_y(tsp_path(j))-original_y(tsp_path(i))]);
  text((original_x(tsp_path(i)) + original_x(tsp_path(j))) / 2,
(original_y(tsp_path(i)) + original_y(tsp_path(j))) / 2,
num2str(len, '%.2f'), 'Color', 'black', 'FontSize', 7);
end
```

```
scatter(original_x, original_y, 10, 'b', 'filled'); hold on;
path = [];
for i = 1:length(tsp_path)-1
  len = norm([original_x(tsp_path(i+1))- original_x(tsp_path(i)),
original_y(tsp_path(i+1))- original_y(tsp_path(i))]);
  steps = ceil(len / NumSteps);
  x_segment = linspace(original_x(tsp_path(i)),
    original_x(tsp_path(i+1)), steps);
  y_segment = linspace(original_y(tsp_path(i)),
    original_y(tsp_path(i+1)), steps);

  for j = 1:steps
    plot(x_segment(1:j), y_segment(1:j), 'Color', [0.1 0.1
    0.1], 'LineWidth', 2); hold on;
    plot(x_segment(j), y_segment(j), 'ro-', 'MarkerFaceColor',
'g', 'MarkerSize', 5, 'MarkerEdgeColor', 'k');
    pause(0.05);
  end
  path = [path; x_segment' y_segment'];
end
num_paths = 5;
path_distances = cell(1, num_paths);
for k = 1:num_paths
  path_matrix = eval(['WP_Path', num2str(k)]);
  path_distances{k} = calculateDistances(path_matrix);
end
tsp_path_distances = calculateDistances([original_x(tsp_
path)', original_y(tsp_path)']);
length_paths = zeros(1, num_paths);
for k = 1:num_paths
  length_paths(k) = sum(path_distances{k});
end
length_tsp_path = sum(tsp_path_distances);
for k = 1:num_paths
  subplot(2, 3, k);
  title(['Possible Path ', num2str(k), ', Total Length= ',
    num2str(length_paths(k))], 'Color', 'black', 'FontSize',
    12, 'Units', 'normalized');
end
subplot(2, 3, 6);
titl.e(['Possible Shortest Path, Total Length=',
num2str(length_tsp_path)], 'Color', 'black', 'FontSize', 12,
'Units', 'normalized');
function distances = calculateDistances(path_matrix)
  num_points = size(path_matrix, 1);
  distances = zeros(num_points-1, 1);
```

```
for i = 1:num_points-1
    dist = norm(path_matrix(i+1, :)-path_matrix(i, :));
    distances(i) = dist;
end

end
```

Appendix 2: MATLAB® Code for Simulation of Path Tracking for an Agricultural Mobile Robot

```
% ----------------------------------------------------------------
% Code title:        Path Tracking Controller with PID
% Author:           Redmond R. Shamshiri, redmond@
                    AdaptiveAgroTech.com
% Last update date: March.01.2024
% ----------------------------------------------------------------

function [path, omega_history, delta_history, v_history,
  theta_history, error_history, speed_error_history] =
  run_robot_simulation()
  function [output, integral_error] = pid_controller(error,
    prev_error, integral_error, dt, Kp, Ki,Kd)
    P = Kp * error;
    integral_error = integral_error + error * dt; I = Ki *
      integral_error;
    if ~isempty(prev_error)
      derivative_error = (error—prev_error) / dt;
    else
      derivative_error = 0;
    end
    D = Kd * derivative_error;
    output = P + I + D;
  end

  waypoints = 0.2*[0 0;30 14;0 35;12 75;83 77;96 15;0 0];
  path = [];
  omega_history = []; delta_history = []; v_history = [];
  theta_history = []; error_history = []; speed_error_history = [];

  L = 1.4;       W = 0.8;       H = 0.6; x = -10; y = 5; theta = 0;
  v_max = 2.7; v_min = 0.5;  v = v_max;
  delta_max = deg2rad(30);
  dt = 0.1; timesteps = 5000; waypoint_index = 1;
  robot_length = L; robot_width = W;
  robot_shape = [
    -robot_length/2, robot_length/2, robot_length/2, -robot_
    length/2, -robot_length/2;
```

```
      -robot_width/2, -robot_width/2, robot_width/2,
        robot_width/2, -robot_width/2
    ];
    wheel_shape = [
      -0.35, -0.35, 0.35, 0.35;
      -0.15, 0.15, 0.15, -0.15
    ];
    Kp = 2.5;          Ki = 0.5;          Kd = 0.05;
    Kp_speed = 0.8; Ki_speed = 0.02;     Kd_speed = 0.05;
      DistTargetSpeed = 0.5;
    integral_error_speed = 0;
    x_label_position = 0.1;
    y_label_position = 0.95;

    for t = 1:timesteps
      x_d = waypoints(waypoint_index, 1);
      y_d = waypoints(waypoint_index, 2);
      theta_d = atan2(y_d-y, x_d-x);
      theta_error = theta_d-theta;
      theta_error = atan2(sin(theta_error), cos(theta_error));
      [delta, ~] = pid_controller(theta_error, [], 0, dt, Kp, Ki,Kd);
      delta = max(-delta_max, min(delta, delta_max));
      omega = v * tan(delta) / L;
      distance_to_waypoint = norm([x_d-x, y_d-y]);
      speed_error = distance_to_waypoint-DistTargetSpeed;
      [v, integral_error_speed] = pid_controller(speed_error,
        [], integral_error_speed, dt, Kp_speed, Ki_speed,
        Kd_speed);
      v = max(v_min, min(v_max, v));
      x = x + v * cos(theta) * dt;
      y = y + v * sin(theta) * dt;
      theta = theta + (v / L) * tan(omega) * dt;
      path = [path; x y];
      omega_history = [omega_history; omega];
      delta_history = [delta_history; delta];
      v_history = [v_history; v];
      theta_history = [theta_history; theta];
      error_history = [error_history; theta_error];
      speed_error_history = [speed_error_history; speed_error];
      clf;

      subplot(3, 2, 1);
      axis equal;
      hold on; grid on;
      for i = 1:size(waypoints, 1)-1
        plot(waypoints(i:i+1, 1), waypoints(i:i+1, 2), 'k-',
          'LineWidth', 1);
      end
      plot(path(:,1), path(:,2), 'bo-.', 'MarkerFaceColor', 'w',
        'MarkerSize', 4, 'LineWidth', 0.1, 'MarkerEdgeColor', 'b');
```

```
R = [cos(theta) -sin(theta); sin(theta) cos(theta)];
robot_shape_rot = R * robot_shape;
patch(x + robot_shape_rot(1,:), y + robot_shape_rot(2,:),
   [1, 0.5490, 0]);

wheel_front_left = [x + L/2 * cos(theta + pi/4), y + L/2 *
   sin(theta + pi/4)];
wheel_front_right = [x + L/2 * cos(theta-pi/4), y + L/2 *
   sin(theta-pi/4)]; wheel_rear_left = [x-L/2 * cos(theta +
   pi/4), y-L/2 * sin(theta + pi/4)];
wheel_rear_right = [x-L/2 * cos(theta-pi/4), y-L/2 *
   sin(theta-pi/4)];
wheel_front_left_rot = R * wheel_shape + repmat(wheel_
   front_left', 1, size(wheel_shape, 2));
wheel_front_right_rot = R * wheel_shape + repmat(wheel_
   front_right', 1, size(wheel_shape, 2));
wheel_rear_left_rot = R * wheel_shape + repmat(wheel_rear_
   left', 1, size(wheel_shape, 2));
wheel_rear_right_rot = R * wheel_shape + repmat(wheel_
   rear_right', 1, size(wheel_shape, 2));
patch(wheel_front_left_rot(1,:), wheel_front_left_
   rot(2,:), 'k');
patch(wheel_front_right_rot(1,:), wheel_front_right_
   rot(2,:), 'k');
patch(wheel_rear_left_rot(1,:), wheel_rear_left_rot(2,:), 'k');
patch(wheel_rear_right_rot(1,:), wheel_rear_right_
   rot(2,:), 'k');

plot(waypoints(:, 1), waypoints(:, 2), 'go', 'MarkerSize',
   10, 'MarkerFaceColor', 'g', 'MarkerEdgeColor', 'k');
for i = 1:size(waypoints, 1)
   text(waypoints(i, 1), waypoints(i, 2), ['('
      num2str(waypoints(i, 1)) ',' num2str(waypoints(i, 2))
      ')'], ...
   'VerticalAlignment', 'bottom', 'HorizontalAlignment',
      'right');
end

title('Adaptive AgroTech Simulation Framework for Path
   Tracking of Agricultural Mobile Robots');
xlabel('X (m)');
ylabel('Y (m)');

desired_path = waypoints(1:waypoint_index, :); robot_
   driven_path = path(1:end-1, :);
error = zeros(size(robot_driven_path, 1), 1);
for i = 1:size(robot_driven_path, 1)
   distances = vecnorm(desired_path-robot_driven_path
      (i, :), 2, 2); error(i) = min(distances);
end
subplot(3, 2, 2); hold on;
grid on;
set(gcf, 'Color', 'w'); ax = gca;
```

```
ax.LineWidth = 0.5;
ax.XColor = 'k';          ax.YColor = 'k';
plot(1:size(error, 1), error, 'mo-.', 'MarkerFaceColor',
  'w', 'MarkerSize', 6, 'LineWidth', 0.1,
  'MarkerEdgeColor', 'm');
title('Distance error between Robot position and the
  desired Waypoint');
xlabel('Time Step');
ylabel('Error (m)');
subplot(3, 2, 3);          hold on;
grid on;
plot(1:t, delta_history, 'bo-.', 'MarkerFaceColor', 'w',
  'MarkerSize', 6, 'LineWidth', 0.1, 'MarkerEdgeColor', 'b');
plot(1:t, omega_history, 'ro-.', 'MarkerFaceColor', 'w',
  'MarkerSize', 6, 'LineWidth', 0.1, 'MarkerEdgeColor', 'r');
title('Steering Angle (\delta) in Blue, and Angular
  Velocity (\omega) in Red');
xlabel('Time Step');
ylabel('\delta (rad) and \omega (rad/s)');
subplot(3, 2, 4);          hold on;
grid on;
plot(1:t, v_history, 'bo-.', 'MarkerFaceColor', 'w',
  'MarkerSize', 6, 'LineWidth', 0.1, 'MarkerEdgeColor', 'b');
title('Linear Speed (v)');
xlabel('Time Step');
ylabel('Linear velocity v (m/s)');
ylim([0, v_max]);
subplot(3, 2, 5); hold on;
grid on;
plot(1:t, theta_history, 'ro-.', 'MarkerFaceColor', 'w',
  'MarkerSize', 6, 'LineWidth', 0.1, 'MarkerEdgeColor', 'r');
title('Orientation (\theta)');
xlabel('Time Step');
ylabel('\theta (rad)');
subplot(3, 2, 6);
hold on;
grid on;
plot(1:t, error_history, 'go-.', 'MarkerFaceColor', 'w',
  'MarkerSize', 6, 'LineWidth', 0.1, 'MarkerEdgeColor', 'g');
title('Orientation Error (\theta_{error})');
xlabel('Time Step');
ylabel('\theta_{error} (rad)');
pause(0.01);
if norm([x-x_d, y-y_d]) < 0.5
  disp(['Waypoint ' num2str(waypoint_index) ' reached!']);
  text(x_label_position, y_label_position, ['Waypoint '
    num2str(waypoint_index) ' reached!'], 'Color', 'black',
    'FontSize', 12, 'Units', 'normalized');

  waypoint_index = waypoint_index + 1;
  if waypoint_index > size(waypoints, 1)
```

```
        disp('All waypoints reached!');
        break;

    end

  end

end
save('robot_simulation_data.mat', 'path', 'omega_history',
  'delta_history', 'v_history', 'theta_history', 'error_
  history', 'speed_error_history');
end
```

Index

0-9

5G, 119, 142–143, 155
360-degree scanning, 10

A

A* algorithm, 23–25, 42, 45, 48, 50, 53, 147, 155, 161–163, 172, 175
Ackermann steering, 157–159, 169
actin, 33, 35
actuation techniques, 120
actuator responses, 164
Adafruit VL53L0X, 15, 18
Adafruit VL53L1X, 15, 18
adaptability, 23, 28, 141, 157–158, 160
adjustable driving modes, 96
adjustable suspension system, 97
advanced features, 97
advantages, 12, 89, 108, 111, 141, 157, 159–160, 169
aerial imagery, 57, 69
AgBot, 1–2, 6, 26, 130, 138, 144
AGCO Corporation, 37
Agisoft Metashape, 68–69
AgJunction, 9
agricultural attachments, 95–96
agricultural fields, 1, 10, 12, 26–27, 32, 36, 51, 89, 119, 140–141, 159–160
agricultural geophysics, 55
agricultural implements, 8, 96
agricultural machinery, 3, 6, 40–41, 46, 95, 97, 103, 171–173
agricultural mobile robots, 1–13, 15, 20–21, 24–27, 29, 32, 36–37, 47, 149, 156–158, 160, 164–165, 169, 171, 181
agricultural organizations, 104
agricultural productivity, 103
agricultural robotics, 38, 49, 53, 59, 107–110, 139, 143–146, 148, 165, 172
agricultural robot market, 37
agricultural robots, 2–3, 10, 32, 38–42, 46, 54, 108, 124, 139, 142–143, 146–147, 152, 155–157
agriculture, 1–3, 9, 13, 26, 32, 37–46, 49–50, 52, 54–55, 60, 85–89, 91, 94–95, 98, 104–107, 109, 111, 118–121, 139–152, 155, 157, 159, 171–173
AgroIntelli Robotti 150D, 126
AgroIntelli Robotti LR, 126
AGXEED AgBot, 1, 2, 130
Aigro Up, 126

AI-hosted smart edge platform, 142
algorithms, 1, 4–5, 10, 19–20, 22–29, 34–36, 42, 44–48, 51–53, 86, 107, 109, 112, 115–116, 118, 141, 141–143, 156, 160, 163–164, 168, 172–173
Amazon Ring Indoor Cam, 15–16
analytics, 36
ant colony optimization, 46, 173
anti-rollover system, 96
Apple Harvesting Robot, 149
ARGoS, 33, 35, 53
Arlo Ultra, 15–16
articulated steering, 90–91, 96, 157, 160, 169, 171
artificial intelligence (ai), 91, 107, 111–112, 164
artificial neural networks (ANNs), 165
automation, 3, 6, 38–41, 47–48, 50–54, 86–88, 90, 112–113, 119, 137, 139, 141–146, 148, 151–153, 155, 157, 171, 173
autonomous charging station, 91, 105
autonomous driving technology, 98
autonomous e-tractors, 105
autonomous field robots, 108, 144, 151, 154
autonomous mobile robots, 3, 8, 35, 47
autonomous mowing, 29, 32, 90, 99, 102
autonomous navigation, 1, 3–6, 8, 10, 19–20, 26, 28, 31, 34, 36–37, 39, 41, 46, 51, 59, 87, 91, 102, 105, 113, 124, 141–142, 152, 163
autonomous robots, 29, 37–38, 48, 50, 56, 125, 173–174
autonomous tool carrier, 126, 131
autonomous tractor corporation, 37
autonomous tractors, 3, 115, 159
autonomous weeding, 125, 134
autonomy, 19, 36, 91, 101, 104–105, 118, 141, 153, 165

B

base_local_planner, 29
batteries, 89, 92–94, 104, 106
battery management system, 93
battery pack, 93, 95–97
battery runtime, 90
battery technology, 89, 93–94, 104
bayesian segmented deep learning, 111
behavior-based navigation, 19, 45
Bellman-Ford Algorithm, 161
Benewake TFmini, 17
bidirectional search, 161

blink outdoor camera, 15–16
blueberry picking, 122
blue river technology, 2, 26, 49
BoniRob, 6, 26, 134, 138, 153
Bosch Deepfield Robotics, 37, 134
boundary configurations, 73, 84
Braitenberg, 23, 27, 47, 50
Branch and Bound, 162–163, 173
bug algorithms, 19, 45
bulk density, 55

C

cable-operated, 120
calibration tests, 72, 80
California-based, 96
cameras, 3, 5–6, 8, 12–16, 25–27, 31–32, 35,
    42–43, 56, 59, 91, 101, 109–112, 114–117, 125,
    129, 134, 136, 141, 143–144, 147–148, 164
camera systems, 12, 134, 137, 139
CANBUS communication, 19, 44, 99, 106, 152, 172
Cartesian coordinates, 29, 51
cartographer, 29–30, 43
Case IH, 3, 8–9, 37, 97–98
charging infrastructure, 93, 103–104
charging times, 104
Christofides algorithm, 162
citrus tree grove, 64, 77, 80, 83
Clearpath Jackal UGV, 57–58
climate change, 92, 107, 118
cloud computing, 35, 142, 145
clustering algorithms, 29, 52
CMD-Tiny meter, 59, 61
CNH Industrial, 3, 6, 37, 97
collaborative robots, 85, 118, 142, 149–150
collision avoidance, 1, 10, 12, 17, 19–24, 26–27,
    31–32, 34, 36, 38, 40, 42–44, 46, 48–49, 102,
    106, 124–125, 141–142, 152, 172–173
commercialization, 36–37, 143–144
communication interfaces, 29
communication modules, 130
complementary filters, 25
complex environments, 19–20, 23–24, 27–28,
    139, 163–164
computational complexity, 20–22
computer vision algorithms, 5, 109
contact pressure, 120
Continental Contadino, 134
continuum arm, 119–120
control algorithms, 26, 141–142, 160, 168
control strategies, 34–35, 48, 117, 120, 147, 150,
    157, 169
control systems, 44, 109, 112, 139, 141
convolutional neural networks (CNNs), 27, 111
CoppeliaSim, 32–34, 53, 157, 165, 167, 169
correlation-based methods, 112

cost, 10, 15, 20, 25, 29, 37, 39–40, 43, 57–58,
    85, 87, 89–92, 94, 96, 103–104, 106, 112,
    119–120, 124, 141, 150, 157, 161, 163, 168
costmap_2d, 30–31
cost reduction, 104
crop management, 144
crop monitoring, 1–3, 26, 38, 43, 119, 127, 135,
    159, 171
crop scouting, 130

D

Dahlia Robotics Dahlia, 3.3-4.3, 132
data acquisition devices, 125, 142
data analysis, 66, 91
data communication, 1, 8, 101
data management, 8, 36
data transfer, 130, 142
DBSCAN (Density-Based Spatial Clustering of
    Applications with Noise), 29
dead reckoning, 5, 28, 39
decision-making models, 142
decision support, 40, 104, 130
deep learning, 31, 39, 49, 52–53, 86, 111–112,
    115–116, 142–143, 147–149, 151
deep neural networks, 27, 148
deformable materials, 119
deliberative planning, 26–27
dense point cloud, 10
depth camera, 30–31, 34, 43, 113, 115, 148
depth information, 10, 13, 109, 115–116
diesel engine, 90, 94, 97, 126, 130
differential correction, 136
differential drive, 51, 157–159, 165, 167, 169, 171
differential high-precision multi-band GNSS, 60
digital agriculture, 109
digital cameras, 109
digital farming, 145–146
digitalization, 1, 40, 90, 104, 109, 143
digitalization of agriculture, 104
digital technology, 107
digital transformation, 142
Dijkstras algorithm, 20–21, 23–24, 143, 161, 163
disadvantages, 89
disease detection, 115, 125, 151, 158
distance detection sensors, 3, 12, 15, 17, 34, 101,
    111, 141
distributed automation systems, 142
disturbances, 24, 32, 165, 168
dual-arm robot, 109
Dwa_local_planner, 29, 31
dynamic environments, 19, 27, 45–46, 163
dynamic obstacles, 10, 21–22, 24, 42, 164
dynamic programming, 162–163, 173
dynamic window approach, 23–24, 27, 31,
    47–48

E

E25, 94–96
eAutoPowr transmission, 97
ECa map, 58, 73, 77
ECa measurements, 57–58, 72–73, 77, 84
eco-friendly farming equipment, 104
economic performance, 118, 149
economic viability, 140
Edison HTZ, 3512 94–95
Edison Motors, 95
efficiency, 1, 3–4, 58, 95, 97, 103–104, 107–108,
    110–111, 116, 118, 142–143, 156–157,
    159–160, 171
electrical and pneumatic circuit, 121
electrical motors, 126–127, 129–137
electrical resistivity methods, 55
electric drive system, 95
electric farm tractors, 103
electric motor, 89, 95–97
electric motor design, 89
electric tractors, 89, 95, 104–105
electromagnetic fields, 63
electromagnetic induction (EMI) sensor, 55, 80
electromagnetic interference, 62–65, 68, 71, 73, 80
electromagnetic interference mitigation, 7
embedded image processing, 115
EMI sensor, 58, 61, 63, 70, 80
emissions reduction, 91
encoder measurements, 115
end-effector, 113–115, 139–140, 144
energy consumption, 20, 118, 163, 173
energy efficiency, 156
environmental modeling, 6
environmental patterns, 8
environment reconstruction, 109
Escarda Technologies Escarda, 135
ESRI ArcGIS, 73, 77
ETAROB, 133
Ethernet, 7–9, 11, 16, 18, 29, 115, 127–128
E-tractor manufacturers, 104
E-tractors, 89–95, 97–99, 101–104
euclidean clustering, 29
euclidean distance, 163
EufyCam 2C, 15–16
exhaust emissions, 89
Exobotic LAND, 127–128, 138
exponential semivariogram, 73–75, 77, 81, 83
extended Kalman Filter (EKF), 20, 60
eye-in-hand configuration, 111, 113, 116
eye-to-hand configuration, 110

F

farm automation, 3
Farmdroid FD20, 159

farmers, 10, 36–37, 89–94, 96–97, 103–104,
    107–108, 118, 124–125, 143, 145, 151, 153, 157
farming practices, 1, 3, 90, 103–104, 108, 144
Farming Revolution farming GT, 129
farming tasks, 108
Farmtrac FT25G, 94, 96–97
feasibility study, 152
feedback mechanisms, 165
Feldschwarm Technologies Feldschwarm, 138
FEM software, 120, 140
Fendt e100 Vario, 90, 96
fertilization, 92, 108, 135, 139
fertilizing, 133–134
field conditions, 1, 8, 15, 32, 56, 61, 101, 107, 118,
    125, 141, 157–158, 169
field monitoring, 157
Field of View (FoV), 10, 16
field robots, 54, 108, 124–125, 139, 141–142, 144,
    151, 153–154
field-scale ECa map, 58
field-scale experiments, 73, 76–77
field survey, 58
fine-tuning, 168
finger-tracking gloves, 122–124
fish-eye cameras, 12, 43
flexibility, 12, 34, 97, 104, 109–110, 114,
    119–120, 142, 160
Floyd-Warshall algorithm, 161
fluidic elastomer actuators (FEAs), 121
follow the gap, 23, 27, 47
force transducer, 120
Fraunhofer IPA CURT, 135
front loader, 95–96
fruit detection, 109, 115, 148, 154
fruit harvesting, 3, 109, 116, 122, 144, 147, 149
fruit localization, 112, 115
fruit picking, 39, 122, 140
fruit visibility, 117, 144
fuel consumption, 93, 104
fuel costs, 90
funding, 37, 56, 104, 107, 143, 145
Fuzzy Logic Control, 165

G

Gazebo, 32–34, 53, 69–70, 88
Gazebo Robotics simulator, 69
genetic algorithm, 46, 48, 161
Gentle Robotics E-Terry, 135
geodetic coordinates, 29, 51
geofencing, 29
Geographic_msgs, 29
geological mapping, 55
geometric adjustment, 144
geometric requirements, 125
georeferenced prescription maps, 114

georeferencing, 29
GIS (Geographic Information System), 29
Global_planner, 29
Global Navigation Satellite Systems (GNSS), 2,
   4–5, 7, 9, 57, 59, 60, 63, 70, 73, 77, 126
global path planning, 20, 22–24, 29, 172
gmapping, 29–30, 51
GNSS, 4–5, 7, 9, 39–41, 57, 59–60, 63, 70, 73,
   77, 136
GNSS-tagged trajectory, 60
Google Nest Cam IQ, 15–16
governments, 37, 103, 107, 143
government support, 94
Gps_common, 29–30
GPS, 3–9, 21–22, 25–31, 35–36, 39, 47, 52, 58,
   60, 73, 77, 88, 92, 99, 101–102, 113, 125–132,
   134–137, 141–142, 155, 164
GPS-based navigation, 8, 101–102, 142
GPS navigation, 9, 30, 99
GPS Receivers, 6, 29
gradient-based optimization, 169
gradient-free optimization, 169
grants, 103–104
grapevine canopy, 115
graph-based algorithms, 163
graph-based SLAM, 21, 23
graphics processing units, 109
graph search algorithms, 163
grasping ability, 119
grasping force test, 122
grasp selection, 117
greenhouse gas emissions, 89
grid-based occupancy mapping, 21, 23
gripping force, 122
gripping process, 144
grip strength, 120
ground mobile robot, 58, 80
ground-mounted solar systems, 103

H

Hürlimann, 97
handling ratio, 121
harvesting, 1–3, 37, 90, 92, 108–112, 114–123,
   138–141, 143–144, 146–150, 154–156, 160,
   171
harvesting movements, 122
harvey sweet pepper harvesting robot, 116
HC-SR04, 15, 18
HC-SR04 ultrasonic sensors, 15
heavy actuation mechanisms, 119
heavy-duty tasks, 94
heavy workloads, 90
herbiciding, 118
heuristic approach, 19, 163
high-accuracy positioning module, 60

higher yields, 124
high-value crops, 119
H-infinity control, 165
Hokuyo URG-04LX, 5, 11, 17–18
Holybro H-RTK F9P GNSS series, 60
Holybro Pixhawk 4 autopilot module, 60
Holybro SiK Telemetry V3, 60
horticultural engineering, 144
human-robot collaboration, 109, 118–119, 144
human workforce, 1, 112, 144
hydrogen fuel cell e-tractors, 95
hydrostatic transmission system, 97
hyperspectral cameras, 12
hyperspectral sensors, 5

I

IBVS (image-based visual servoing), 109–110,
   112, 117
image-based localization, 8
image-based visual servoing (IBVS), 110
image processing, 109, 115, 139–140
imaging sensors, 109
impact forces, 122
IMU (inertial measurement unit), 5, 141
incentives, 103–104
induction motor, 96
industrial robots, 35, 49, 119
inertial measurement units (IMUs), 5, 141
infinite degrees of freedom, 120
infrared cameras, 12, 136
infrared Sensors, 22–23, 44, 106, 152, 172
infrastructure investment, 94
inherent flexibility, 120
Innok Robotics HEROS
inspection, 56, 61, 71, 80, 86, 109, 159
Intel D435i Stereo Camera, 14
intelligent EMS, 95
intelligent energy management system
   (IEMS), 96
intelligent systems, 38, 50, 54, 144, 152,
   172
Intel RealSense D435, 13–14, 17–18, 109
Intel RealSense D455, 14
Intel Realsense T265, 13–14
interface, 6–7, 9, 11, 16, 18, 34–35, 69, 97,
   100–101, 118
Internet-of-Things (IoT), 104
inverse kinematic mode, 112
inverse kinematics, 53, 115
IoT-based cloud computing, 142
IP62 rugged robot, 61
irrigation scheduling, 58
IR sensors, 6, 15
ISO-BUS protocol, 99
IT infrastructure, 144

J

Jackal robot, 59–61, 68–69, 72
JeVois-A33 Smart Camera, 14
John Deere, 2, 8–9, 37, 91, 93–95, 97, 105, 136
John Deere SESAM, 93–95
joint angles, 113
joint velocity commands, 115

K

Kalman filter, 20–21, 25–26, 45, 47, 49, 51, 60
Kalman filtering, 47, 165, 174
kinematic equations, 165, 168
Kinova Gen3, 114–115
Kioti Mechron 2240 Electric, 97–98
Kramer 5055e, 94–95
Kramer-Werke GmbH, 95
kriging interpolation, 73–75, 77, 80–81, 83
Kubota Corporation, 37
Kubota X tractor, 97–98

L

labor costs, 107
ladybird, 26, 49, 159, 172
large-scale agricultural operations, 104, 158, 160
laser beams, 6, 10, 17
LaserScan, 31, 70
laser weeding, 132
latitude, longitude, altitude, 29
lead-acid batteries, 89
LIDAR, 3, 5–6, 8, 10, 12, 15, 17, 21–23, 25–28, 30–31, 34, 38, 40, 42, 46, 49–52, 59, 63, 101, 113–114, 125, 152, 164
LiDAR-IMU fusion, 25
LiDAR sensors, 10
lifting capacity, 126–137
Li-ion battery, 126
linearity, 58, 65, 68, 73, 80
Linear Quadratic Regulator (LQR), 165
linear regression, 65–66
LiPO4 battery, 99
lithium-ion batteries, 94, 95–97, 106
live monitoring, 104
Livox Mid-40, 17
localization, 4, 6, 8, 10, 20–23, 26, 29–30, 34, 39–41, 44–47, 49, 51–52, 57, 84, 88, 111–112, 115–116, 139, 148, 155, 174
localization and mapping, 6, 10, 20, 29, 40–41, 52, 155
local path planning (LPP), 24
LoRaWAN, 119, 142
Lorex Security Cameras, 15
lower costs, 121

lower operating costs, 92
LQR (Linear Quadratic Regulator), 25, 165

M

machine learning, 2, 4, 27, 31, 36, 50, 57, 107, 109, 140, 142, 145, 164
machine learning algorithms, 36, 107
machine vision, 39, 107, 117, 144–145, 149
machine vision systems, 107
maintenance, 6, 91–92, 94, 104, 128, 141, 157, 159–160
maintenance costs, 92, 104
maneuverability, 73, 92, 96–97, 157–158, 160
manipulation study, 122
manipulator tasks, 119
manual baselines, 73, 77
mapping, 1, 4, 6–7, 10, 13–14, 17, 20–23, 26, 29–30, 34, 36, 39–43, 47, 51–52, 55, 58–59, 80, 86, 120, 136, 151, 155, 171
maps, 5, 8, 10, 15, 23, 28–31, 69, 73–75, 77, 80–81, 83, 114, 154
market availability, 94, 124
market fluctuations, 107
MATLAB, 34, 53, 112, 140, 148, 157, 164–165, 168–169, 175
MAVROS software, 60
MaxBotix, 18
measurement inaccuracies, 164
measurement linearity, 65
mechanical damage, 122, 151
mechanical weeding, 49, 124, 129
medical and industrial fields, 119
Microsoft Azure Kinect DK, 14, 43
minerals content, 55
miniaturization of sensors, 107
mobile platforms, 108, 114, 152
mobile robotics, 38, 41–42, 47, 146, 151, 160, 163
mobile robots, 1–10, 12, 14–15, 19–21, 24–32, 34–39, 42, 44–48, 50–51, 53–54, 56, 125, 141, 146, 149, 156–160, 163–165, 169, 171–173, 181
models, 4, 15, 20, 27–28, 31, 34–35, 37, 61, 69, 89–92, 94, 96, 101, 103–104, 140, 142, 157, 168
modular design, 127
modularity, 141
moisture meters, 55
monarch e-tractor, 97
monitoring, 1–3, 8, 12–13, 26, 28, 38, 40–44, 51–52, 85–86, 94, 97, 104, 119, 127, 130, 132–136, 139, 155, 157, 159, 166, 171
Move_base package

mowing, 2, 8, 15, 17, 29, 32, 89–90, 99, 102, 109, 124, 142
MPC (model predictive control), 20, 22, 25, 157, 165, 168–170
MPC controller, 168–170
muddy terrain, 80
multi-channel distance detection sensors, 101, 141
multi-joint mechanisms, 121
multi-modal belief, 61
multi-purpose agricultural mobile platform, 119
multi-robot arms, 109–110
multispectral, 26, 56, 115, 125
multi-spectral cameras, 12, 141
multispectral imaging, 115
multi-target depth ranging, 115

N

Naio Technologies, 6, 37, 131
Navfn, 29
navigation, 1, 3–6, 8–10, 12–13, 19–20, 23–24, 26–31, 34, 36–46, 50–51, 53, 59–61, 87, 91–92, 99, 101–102, 105, 113, 115, 124–125, 141–142, 145, 152–153, 155, 157, 160, 163–165, 169, 171–172
navigation controllers, 164
navigation systems, 1, 5, 40, 165
Navsat_transform_node, 60
nearest neighbor algorithm, 164, 175
near-optimal solutions, 161–162
net-metering initiatives, 103
New Holland T4, 97
New Holland T6, 90, 97, 106
Nmea_navsat_driver, 29
NMEA sentence parsing, 29
noise pollution, 92
non-invasive sensing techniques, 55
nonlinear dynamics, 164, 168–169

O

object detection algorithms, 116
object identification, 109
observer design, 165
obstacle avoidance, 4, 24–25, 29, 44–45, 47, 50, 61, 143
obstacle detection, 1, 4, 6, 10, 12–15, 17, 26, 29–31, 38, 42, 52
occluded fruits, 118
occlusions, 10, 111, 114
occupancy grid maps, 31
odometry, 5, 8, 21–22, 25–26, 30–31, 39, 41, 49, 60, 155, 164
olive tree grove, 63–64, 77–78
onboard sensors, 56, 59, 61, 63

on-site charging stations, 103
OpenCV, 14–15
open-source image processing software, 109
open-world navigation, 61
operating times, 104
operational range, 93
optical sensors, 114
optimal fruit ripening, 119
optimal navigation, 19, 164
optimal sensor placement, 62, 72
optimization algorithms, 162
Optitrack cameras, 115
orchard, 2, 37–38, 48, 77, 85, 99, 101, 104, 108, 112, 114, 130, 140, 142, 146, 153–154
organic farming, 153
Ouster OS-1, 11

P

panoramic perspective, 10
Pan-Tilt-Zoom (PTZ) cameras, 12–13, 15
Parallax PING))), 15, 18
particle filter, 21, 23, 25, 34, 45, 51
path length, 163
path planning, 8, 10, 20–24, 29–31, 34–35, 39–42, 44–48, 50–51, 53, 108, 115, 124, 143, 147, 149, 163–164, 172–173
path tracking, 20–25, 46, 48, 156–157, 164–171, 173, 179, 181
path tracking performance, 157, 165
payload capacity, 59, 96, 98
Pcl_ros, 31
PCL (Point Cloud Library), 31
Pearson correlation, 65–66, 73, 80
Pearson correlation coefficient, 73, 80
peduncle segmentation, 117
performance evaluation, 40–41, 147, 153
pesticide applications, 115
pest management, 159
phenotyping, 26, 28–29, 38, 49–51, 85–87, 108, 114, 134, 146, 153–154, 159, 171
photogrammetric software, 69
physical sampling, 80
physics engine, 32, 35
picking patterns, 122, 124, 144
PID controller, 157, 166–167, 169
plant development, 144
planting techniques, 109
plant phenotyping, 26, 29, 51, 146, 153, 159, 171
Pointcloud_to_laserscan, 31
point cloud, 10, 31, 52, 173
PointCloud2, 31
polynomial fits, 77, 80
potato cultivation, 125, 138–139, 144, 153
potential fields, 19, 22, 27
power consumption, 7, 61–62, 99, 118, 121

precise control, 104, 157, 165
precise maneuvers, 108
precision, 2–3, 7–9, 15, 20, 25–26, 38–42, 46, 52, 56, 60, 85–86, 88, 92, 98, 114, 121, 127, 132, 137, 139, 141, 145–146, 150–151, 154–155, 158–160, 171–172
precision agriculture, 3, 26, 38–39, 41–42, 52, 60, 85–86, 88, 98, 139, 145–146, 151, 155, 159, 171
precision farming, 7, 9, 40, 86, 92, 171
precision spraying, 26, 127, 132, 159
prediction horizon, 168–169
pressure-mapping sensors, 120
price range, 6–7, 11, 14, 16, 18
probe distance, 68, 72–73
production costs, 10, 105, 108, 112, 125
productivity, 1, 3, 89, 92, 96–97, 103–104, 118, 143
proof-of-concept, 104, 122, 142
Prospero, 26
proximal sampling, 80
pruning, 109, 112–113, 118, 120, 172
pruning and thinning, 120
pure pursuit, 4, 22, 25, 46, 165, 172
PVC tubes, 62
PyRobotSim, 33, 35
PyTorch, 31, 52

Q

Qbrobotics, 121
Q-learning, 28, 50
quadratic traversal algorithm, 47, 115, 149

R

RADAR, 5, 22, 26, 136
range, 1, 5–7, 10–12, 14–18, 21, 32, 43, 63, 67, 89–90, 93, 95–99, 101, 104, 106, 108, 110, 117, 125, 139–140, 151, 166
rapidly-exploring random trees (RRT), 20, 24, 34, 163
R-CNN neural network, 115
reactive control, 26–27
RealSense D435i, 115
real-time constraints, 24, 164
real-time control, 35, 143
real-time kinematic (RTK), 2, 7, 60
real-time performance, 164
real-time recognition, 115
rear hitch, 95–96, 100
rear PTO, 96
redundant manipulators, 109
reference maps, 8, 15
regulatory framework, 36
reinforcement learning (RL), 28, 45, 48, 50, 87, 164–165

reliability, 8, 10, 15, 36, 40, 93, 97, 102, 108, 140, 142, 144, 150–151
remote control, 35, 127
renewable energy sources, 92–93, 96
Reolink Argus PT, 15
re-ridging, 138
research and development, 1, 19, 37–38, 93, 104, 107–108, 117, 141, 144, 146
resource-constrained platforms, 164
RGB and RGB-D Cameras, 13
RGB-D camera, 14, 21–22, 30–31, 52, 111, 115–116, 151
RGB images, 28, 115–116
ridge structure, 125, 138
ridging, 126, 138
rigid components, 119
rigid link, 119
Rigitrac SKE 50, 94–95, 97
Ring Stick Up Cam, 15–16
RoboDK, 33, 35, 53
robot_localization package, 29
robot acceptance, 146
robot arm, 110–111, 114–116, 139, 150
robot design, 5, 37
robotic arm, 34, 118–119
robotic harvesting, 111, 114–117, 120, 140, 147–148, 154
robotic manipulator, 109, 120, 122, 149–150
robotic manipulator control, 109
robotic manipulators, 34, 109, 112–113, 119, 124, 139, 144, 150
robotic platforms, 1, 6, 49, 117
robotics applications, 7, 164–165
robotic technology, 139, 142, 144
robotic weeding, 114, 172
robot navigation, 30, 39, 42, 45–46, 50, 87, 145, 155, 163, 171
Robot Operating System (ROS), 28, 112
robot-sensor configurations, 63
robot simulation, 32, 35
robot traversability, 62, 68, 80
robust control, 165
robustness, 8, 28–29, 70, 80, 107, 144, 168
rooftop solar installations, 103
ROS-based vehicle control unit, 102
ROS drivers, 115
ROS libraries, 29, 70
ROS node, 29, 52
rotovating, 126
route planning, 4, 38, 50
RPLidar A2M8, 17–18
RRT, 20–21, 24, 34, 45, 87, 163, 172
RTK, 4–5, 7–9, 40–41, 57–58, 60, 70, 73, 77, 113, 141
RTK-DGPS, 134
RTK-GNSS, 70, 73, 77

RTK-GPS, 8, 113, 141
rural charging infrastructure, 104

S

safety cab, 96
safety features, 5, 15, 96
satellite imagery, 73, 77
scan rate, 11
SDF (Simulation Description Format), 32
seeding, 2, 37, 91, 126, 129–131, 134, 158–160
sensing solutions, 101, 108, 141, 144
sensor_msgs, 31, 70
sensor displacement, 71
sensor examples and specifications
sensor fusion, 8, 20–23, 25–26, 28–30, 34, 41–42,
    47, 49, 155
sensor fusion algorithms, 25, 29
sensor height optimization
sensor measurements, 20, 23, 57, 164
sensor noise, 21–23
sensor oscillations, 70
sensor placement, 58, 62, 68, 70, 72
sensors, 1, 3–6, 8, 10, 12–13, 15, 17, 19, 21–23,
    25–32, 34–36, 40, 42–44, 46, 49, 51–53,
    55–56, 59–61, 63, 86, 101–102, 104, 106–107,
    109, 111, 114–115, 117, 120, 124–137, 141–
    142, 145, 148, 150, 152, 159, 164, 172–173
sensor stability, 68
Sharp GP2Y0A21YK0F, 15, 18
shortest path, 24, 48, 115, 156–157, 160–161,
    163–164, 169–170, 172, 175–177
shortest path planning, 115
SICK LMS400, 11
SICK LMS511, 11, 17–18
signal correction methods, 8
simpler structures, 121
simulated annealing, 162, 173
simulation, 32–35, 45, 47, 50, 53–54, 58, 62,
    68, 72, 101–102, 112–113, 118,
    148, 152, 156–157,
    165–169, 181
simulation environment, 34–35, 68, 118
simulation framework, 156–157, 165–166, 181
simulation software, 32, 53, 148
Simultaneous Localization and Mapping
    (SLAM), 10, 14, 21, 23, 28–31
Single Shot Multibox Detector (SSD), 116
site-specific applications, 124
site-specific spraying, 119
Sitia Trektor, 128
SLAM (simultaneous localization and mapping),
    10, 14, 21, 23, 28–31
Sliding Mode Control (SMC), 165, 168
small farms, 96
Small Robot Company Dick, 132

Small Robot Company Harry, 137
Small Robot Company Tom, 132
smart farming, 37, 40, 145
social acceptance, 119
soft gripper, 121–122, 124, 150
soft grippers, 108, 119, 121, 140–141, 150
soft robotic arms, 119–120
soft robotics, 108, 119–121, 140–141, 150, 154
soft-robotic technology, 144
soft robot manipulator, 119
software architecture, 144
soil apparent electrical conductivity (ECa),
    55–56, 58, 61, 80
soil characteristics, 55
soil compaction, 114, 163
soil conductivity measurements, 59, 61, 63, 65, 80
soil cultivation, 108
soil health, 56, 87, 92, 104, 163
soil moisture measurements, 56–57
soil-plant-management relationships, 55
soil salinity, 55–56, 65, 86
soil sampling, 57, 86
soil sensing, 29, 51, 171
soil spatial variability, 55
soil temperature, 55
solar panels, 89–90, 93–94, 96, 98, 103
solar powered, 129, 133–135
Solectrac, 89, 94–96
specification, 7, 100
speed control, 92, 99, 165–166
spraying, 1–2, 8, 26, 89–90, 109, 114–115,
    118–119, 124, 126–128, 132, 134, 136, 139,
    142, 149, 151, 159, 171
spraying robots, 115
stability, 62, 68, 93, 97, 113, 142, 158, 160, 165,
    168
stage, 4, 33, 35, 117–118, 126, 130, 138
standardized interfaces, 36
Stanley controller, 22, 25, 46
state estimation, 25–26, 60, 165
static calibration tests, 80
steering angle, 24–25, 100, 102, 157, 165–166,
    168–169, 182
steering mechanisms, 156–157, 159–160, 169,
    171
steering systems, 157
Stereolabs ZED 2 Stereo Camera, 14
Stereolabs ZED Mini Stereo Camera, 14
Stereovision, 12, 42
stereo vision, 13–15, 172
straight-line trajectory, 72–73
Strubes BlueBob, 137
subsidies, 103–104
sustainability, 1, 37, 48, 56, 89, 97, 105, 143,
    149–150
sustainable farming, 3, 103, 105, 124

sustainable farming practices, 3, 103
swarm robotics, 35, 54
system dynamics, 20–21, 168–169

T

target spraying, 109, 114–115, 118
task and motion framework, 80
task planning algorithms, 109
tax incentives, 104
technological challenges, 36
technological innovation, 94
technological robustness, 144
teleoperation, 54, 58, 104, 118, 163
teleoperation control, 104
tensorflow, 31, 52
terrain traversability, 62, 68
terrain traversability capacity, 68
thermal cameras, 5, 12
thinning, 109, 112–113, 120
thorvald, 1–2, 6, 26, 38, 159, 172
three-point hitch system, 96
Time of Flight (ToF), 5–6, 134
timestamp information, 29
topology-optimized gripper, 121
touchscreen interface, 97
TP-Link Kasa Spot, 15–16
tractor, 3, 6, 19–20, 37, 41, 46, 50, 89–106, 130,
     160, 171, 173
trajectory generation, 4, 24, 38
traveling salesman problem (TSP), 162
traversability, 58, 62, 68, 80
tree shapes, 109
trimming, 109, 112–114
trimming robots, 149

U

UART, 7, 29
ultrasonic, 5–6, 12, 15, 18, 23, 30
ultrasonic sensors, 6, 12, 15, 30
uncertain environments, 164
Unity3D, 33, 35, 53
Universal Transverse Mercator (UTM), 60
Unmanned Aerial Vehicles (UAVs), 56
unmanned machinery, 142
Unscented Kalman Filter (UKF), 22, 25, 29
unstructured environments, 10, 110, 119, 121,
     141, 144
UR3 robot, 116–117
UR5 robot, 114, 116
USB, 7, 11, 18, 29, 60, 117
U-shaped trajectory, 72–73, 75, 77

V

validation experiments, 72
Velodyne, 5, 11, 17–18, 114
Velodyne VLP-16, 5, 17
venture capital, 37
versatile robotic systems
versatility, 95
vertical oscillations, 70
virtual environments, 32, 35, 53, 102, 148
virtual replicas, 104, 116, 142
vision-based control, 112–113, 116, 147, 149
vision-based navigation, 39, 142
vision-based sensors, 5, 12
visual features, 8, 10, 110
Visual-Inertial Odometry (VIO), 25–26
visual odometry, 8, 39, 41
visual servoing, 109–112, 115, 118, 144, 147–148

W

warning signals, 122
wavelength, 12
waypoint-based path tracking, 156
waypoints, 24, 29, 58, 72, 156–157, 163, 165,
     168–170, 179–182
weather conditions, 2, 5, 10, 15, 27, 36, 56, 104,
     108
Webots, 32, 35, 53
weed control, 109, 118, 134–135, 138–139, 151
weed detection, 114, 127, 132, 159
weeding, 2–3, 8, 26, 47, 49, 114–115, 118,
     124–127, 129–132, 134–135, 137–139, 142,
     145, 149, 152–153, 157–159, 172
weeding robot, 3, 115, 127, 153, 158
Weedingtech, 97–98
Weidemann 1160 eHoftrac, 96
weight, 90, 93, 99–100, 116, 121, 124, 160–161
wheel encoders, 8, 29, 59–60
wind turbines, 103
wireless communication, 104, 107
wireless teleoperation, 163
workforce decline
workstation PC, 115
Wyze Cam V3, 15

X

XML, 32

Z

Zauberzeug Field Friend, 133